Contouring
Geologic Surfaces
with the Computer

COMPUTER METHODS IN THE GEOSCIENCES

Daniel F. Merriam, Series Editor

Computer Applications in Petroleum Geology
Joseph E. Robinson
Graphic Display of Two- and Three-Dimensional Markov
  Computer Models in Geology
Cunshan Lin and John W. Harbaugh
Image Processing of Geological Data
Andrea G. Fabbri
Contouring Geologic Surfaces with the Computer
Thomas A. Jones, David E. Hamilton, and Carlton R. Johnson
Exploration-Geochemical Data Analysis with the IBM PC
George S. Koch, Jr.

Related Titles

Statistical Analysis in Geology
John M. Cubitt and Stephen Henley (eds.)
Cluster Analysis for Researchers
H. Charles Romberg
Analysis of Messy Data, Volume 1: Designed Experiments
George A. Milliken and Dallas E. Johnson

CONTOURING
GEOLOGIC
SURFACES
WITH THE
COMPUTER

THOMAS A. JONES
DAVID E. HAMILTON
CARLTON R. JOHNSON
Exxon Production Research Company

VAN NOSTRAND REINHOLD
——————————— New York

Manufactured in the United States of America.

Published by Van Nostrand Reinhold
115 Fifth Avenue
New York, New York 10003

Van Nostrand Reinhold International Company Limited
11 New Fetter Lane
London EC4P 4EE, England

Van Nostrand Reinhold
480 La Trobe Street
Melbourne, Victoria 3000, Australia

Nelson Canada
1120 Birchmount Road
Scarborough, Ontario MIK 5G4, Canada

15 14 13 12 11 10 9 8 7 6 5 4

QE
36
.J66
1986

13394263

7-19-91 Ac

Library of Congress Cataloging-in-Publication Data
Jones, Thomas A., 1942-
  Contouring geologic surfaces with the computer.
  (Computer methods in the geosciences)
  Includes index.
  1. Geological mapping—Data processing. I. Hamilton, David E.,
1952-   II. Johnson, Carlton R., 1926-   III. Title. IV. Series.
QE36.J66   1986   550   86-7776
ISBN 0-442-24437-1

Contents

v

12 Historical Reconstruction   255

Afterword   273

Appendix A: Program Capabilities   277

Appendix B: Stratigraphic Example   289

References   299

Index   309

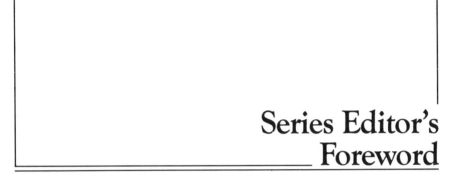

Series Editor's Foreword

Graphics are the end product of any analysis, whether as a listing or multicolored computer CRT display, and are the aspect of any project that probably receives the least attention from the analyst. Graphics, of course, are important because of the influence on the interpretation depending on the presentation. Although graphics take many forms, contouring two-dimensional surfaces is one of the most important and widely used techniques used by geologists and other earth scientists. It is in this manner that three-dimensional aspects can be represented in two-dimensional space. Combine with the computer the various methods of contouring and the representation of results by this form, and you have a powerful tool to analyze data.

The development and implementation of software generally has been dependent on the development of hardware. Because of the small but powerful microcomputer and many software packages developed both in academia and industry, mapping software is now available. Consequently, there is a need for guidance in the use of different mapping methods. *Contouring Geologic Surfaces with the Computer* provides the background desirable for those interested in spatial analysis, including information on the data, methods, applications, and interpretation — all of a practical nature.

As the authors note, the book is not meant to be a user's manual but a supplement to acquaint the reader with the subject. Method coverage is inclusive: introduction, concepts, data, gridding, displays, and applications. Geological coverage concentrates on mapping structural surfaces and thickness intervals. The authors justify their contribution by outlining the advantages of the use of computers for mapping: (1) for manipulating large quantities of data; (2) for speed; (3) for manipulating maps; (4) for updating; (5) for objectivity and consistency; and (6) for incorporating geologic interpretations.

Contouring Geologic Surfaces with the Computer fulfills the objectives of the series well. It is written by geologists for geologists, the topic is of considerable interest, and little is available in one place on the subject. The value of this contribution lies in its practicality—the authors routinely use what they present here—and that use supplements and enforces the geologist's experience. Again, as noted by the authors, "Geological interpretation and the computer cannot be treated separately, but must be closely interwoven." They demonstrate this symbiotic relationship well.

The book can be used by students and practitioners alike. It may be a primary source of information for some, a text and supplement, and a reference for others. It is appropriate for regular course instruction, short courses, or self-study. With the fast moving developments of the field, this book is "must" reading for all those interested and involved in computer applications to solving geological problems.

D. F. MERRIAM

Preface

Computers started becoming generally available to earth scientists during the early 1960s. Since then, the use of computers has expanded manyfold, both in academia and in industry. Indeed, attendees at recent national geological conventions cannot fail to have noticed the many vendors exhibiting computing equipment and programs. The generation of contour maps is a major application of such products. Paralleling the development of the computer, contour-mapping packages have grown from primitive programs to complex integrated systems.

We continue to observe many situations in which the computer and mapping program are used as a "black box." Data are passed to the machine without special instructions, and the resulting maps are accepted without question, even though the maps ignore (or even violate) geologic knowledge or interpretation. Modern sophisticated programs have the capability to incorporate geologic information, and geologists are negligent if that information is not used.

In this book, we stress that two types of information are available for the typical mapping project. The first type consists of the well-known data points. However, the second type of information — often ignored or not recognized as such — consists of geologic knowledge, principles, and interpretation. Maps that are constructed without taking this important information into account could be in error.

The purpose of this book is to introduce tools or methods that allow a geologist to use a mapping program so that it incorporates geological interpretation during contour mapping. Many geological variables are mapped, but applications that involve structural or stratigraphic surfaces continue to dominate. In this book we therefore concentrate on methods for mapping structural surfaces and thickness of intervals between such surfaces.

Application of these methods can prevent such unreasonable results as structure maps containing contours where the surface is known to be missing due to truncation or nondeposition, faults with inconsistent throw, faults treated as folds, or thickness maps with unrealistic, wandering zero contours. These methods also can aid in calculating reserves or correcting mis-ties in geophysical surveys.

The methods we present are just that—general methods to be applied with the program the geologist has available. The program need not be specially tailored because most modern mapping systems can be used without reprogramming. Each of these methods is essentially a series of steps using program options that allow introduction and modeling of geologic interpretation.

Our techniques are not tied to a specific program; they are generic. They are not discussed in terms of specific program parameters, but in terms of general capabilities. Neither are the methods limited in terms of style of operation; they are applicable for interactive processing or batch-submitted jobs, for micros or mainframes.

Our intended reader is a geologist with access to a modern mapping program and enough experience to be able to enter data and obtain simple maps. Such a person can immediately use the methods discussed here. This book is not a replacement for the program user's manual, but is meant to be a supplement.

The book essentially consists of three parts. The first part covers introductory and review material, data entry, and simple contour mapping. The second part is concerned with special data considerations, faulting, and mapping sets of structural-stratigraphic horizons. The third part discusses miscellaneous applications and extensions of the previous concepts, including trends and bias, volumetrics, and construction of paleocontour maps. Appendixes describe required program capabilities and present a set of example data.

ACKNOWLEDGEMENTS

This book is based on research at Exxon Production Research Company during the past 15 years, although Exxon does not endorse or sponsor the book. We thank Exxon for permission to publish this material; special thanks are due to Al Rogers and George Thomas for extra effort on our behalf. Exxon also furnished drafting services.

Bill Byrd, Jim Downing, Steve Hunt, Chuck Iglehart, Kevin McCarthy, Kermit Graf, and Robert Palmquist read all or part of the manuscript and made many helpful suggestions.

We also thank our many colleagues, past and present, who contributed to our knowledge of computer mapping, either through development of Exxon's mapping program or by discovering processing techniques. Among many others, these include Bill Byrd, Henry Dumay, Doug Fenton, Ian Grierson, Dan Hayba, Ron Herdman, Mona Hobby, Neal Jordan, Joe Klingshirn, Mark McElroy, Danny Phillips, Brian Proett, Bill Ryman, and Bob Wilbur.

We also thank Toby Turner, Bill Tippit, and Doug Palkowsky for providing encouragement, aid, and support.

THOMAS A. JONES
DAVID E. HAMILTON
CARLTON R. JOHNSON

Contouring
Geologic Surfaces
with the Computer

CHAPTER 1

Computers, Contouring, and Geologic Interpretation

INTRODUCTION TO COMPUTER MAPPING

Geologists routinely perform three-dimensional analyses to understand and describe spatial relationships, but the typical mode of display — maps and cross sections — is two-dimensional. Contour maps are useful for this analysis, and as soon as the computer became available to the geologic community in the 1960s, interest arose in generating such maps by computers. The early programs were primitive, but they showed that the computer could be a valuable mapping tool.

Rapid improvements in storage capacity and computational speed enabled these primitive programs to grow into sophisticated mapping systems. At present, many contour-mapping programs, with different levels of capability, are available for many machines, from mainframes to microcomputers. This abundance of programs would not exist unless substantial advantages were associated with computer mapping.

Advantages of Computers for Mapping

Geologists have determined computers to be valuable mapping tools because modern programs offer important capabilities.

Manipulate large quantities of data. Modern technology allows measurements

1

on many variables to be collected quickly, providing numerous data points. It is difficult to handle so much information, and the ability to store and manipulate large amounts of information is an important aspect of computer mapping.

Speed. The computer can contour data more rapidly than can a geologist, especially when a series of maps must be constructed. Creating several maps with the computer requires little more effort than creating one, whereas the geologist might require an equal additional amount of time to draw each map.

Manipulate maps. It may be necessary to manipulate old maps to create new ones. For example, one structure map may be subtracted from another to create a map of thickness of the intervening interval. This process requires tedious cross-contouring by hand, but the computer can operate quickly on two surfaces to create a third.

Updates with little effort. Easy updates are an important advantage; if additional data are obtained after a map is created, an update requires that the steps done previously be repeated. With the computer, virtually no effort is required, but manual updating can require as much time as did the original map. Even if only a few points are added, computer processing has an advantage. A geologist with an existing map typically modifies it locally to accommodate the new data. As time goes on, patching new points into the old interpretation can degrade both the appearance and the geological reality of the map. However, independent analysis of the combined old and new points could lead to a new interpretation and major revision.

Objective, consistent maps. Studies comparing computer- and hand-drawn maps indicate that well-designed programs draw maps that are reproducible, consistent, and objective. Dahlberg (1972, 1975) gave subsets of data points from known areas to experienced geologists. The points were contoured both manually and by computer, and the maps were then compared to the known geology. The geologists differed in accuracy, and the computer was at an intermediate position. Dahlberg determined that the computer-generated maps were basically an average of the various geologists' interpretations; he concluded that computer mapping is objective. Modern programs allow the user great flexibility in controlling the appearance of the map.

Incorporate geologic interpretation. Stratigraphic relationships (e.g., truncation) in a region may be incorporated through generation and mapping of subcrop lines. Procedures described in this book allow the incorporation of geologic interpretations into objective computer-drawn maps.

A potential disadvantage of computer mapping is the data-hungry nature of the computer. Best maps are obtained if many data points are available. Further, unreasonable extrapolations in areas of poor data control are possible. However, even with small data sets or if the advantages listed above do not apply, when data are in digital form it is usually worthwhile to create maps with the computer, if only to give a quick look at the data.

Studies Appropriate for Computer Mapping

Not all projects are appropriate for computer usage, and some may profitably be done by hand. However, projects with the following characteristics are recommended for computer processing.

Large data sets. Projects involving large data sets, with several thousand points, can only be done efficiently with the computer. This mapping can range from geophysical surveys in frontier areas to regional studies in densely drilled areas.

Computer-accessible data. If the information is already in a computer database, the project is a good candidate for the computer. Much time in mapping projects is taken up with data gathering and handling. With the information in an accessible form, the geologist can concentrate on mapping.

Updates. Experienced computer users know that data gathering, organization, and input can require a large effort. Similarly, putting together program instructions can be time consuming. In fact, it is not unusual for a first-time study to take longer when it is done by computer than when it is done by hand. However, subsequent computer work with additional data may be finished more quickly than with manual methods. If program instructions are saved, later repeats with added or updated information can be done more rapidly than the original study. Similarly, minor modifications to the original process rarely add substantial time to completion. In short, an appropriate project for the computer is one in which updates are expected.

Need to test multiple hypotheses. Another aspect of repeated mapping occurs when multiple hypotheses are tested. For instance, faulting may be known to exist in a region, but it may be possible to interpret the faults in many ways. When mapping is done by hand, time constraints typically force selection of the one interpretation that seems most likely, and the map is drawn accordingly. When the computer is used, the basic dataset remains the same, but the program instructions change from situation to situation. Testing four hypotheses with the computer may take no longer than testing one by hand.

Need for analysis by other programs. Extensive manipulation or analysis may be done on mapping results, usually by using other programs to analyze the computer's representation of the maps. For example, special programs using computer maps speed the difficult task of mine planning and scheduling. Estimation of petroleum reserves, reservoir simulation, and statistical analysis are other uses of mapping results.

INCORPORATING GEOLOGIC INTERPRETATION

In the previous section, we pointed out the objective nature of computer maps as well as the ability to test multiple hypotheses. These features are not contradictory, as geologic interpretation can be introduced by various methods. In fact, the main theme throughout this book is the inclusion of geologic interpretation into structural and stratigraphic mapping.

Both geologists and computer scientists tend to ignore interpretation in computer mapping, possibly because it is difficult to quantify. Anyone who expects a computer-mapping program to operate effectively when only measured data is supplied will be disappointed. When this happens, the geologist usually believes that computer mapping is not useful, when in reality the lack of a strong and well-integrated geologic interpretation is responsible for a set of incorrect maps. If the geologist does not operate the program, it is his or her responsibility to provide interpretation to the mapmaker and to understand the mapping process. Geologic interpretation and the computer cannot be treated separately, but must be closely integrated.

Types of Projects

A wide variety of projects, ranging from simple to complex, are often handled by the computer. A simple project involves one or more surfaces that do not intersect one another. Complex projects involve several surfaces, some of which intersect one another. The computer-mapping procedures used for simple and complex projects are different. However, modern complete mapping systems available today have capabilities to handle both types.

Simple single-surface projects assume no other horizons exist, as shown by the cross section in Figure 1.1A. This is clearly an unlikely assumption, as such an isolated surface is rare. Mapping the data directly is usually all that is required. However, special processing might be required to incorporate such interpretation as trends or faults into the map. This type of project includes mapping such geologic attributes as geochemical content, rock density, porosity, magnetic field strength, and many others.

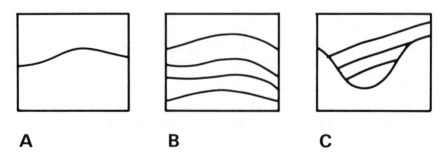

A   **B**   **C**

Figure 1.1   Cross section diagramatically showing the major types of surface mapping projects. (A) Simple single-surface: one independent surface. (B) Simple multiple-surface: several conformable or nonintersecting surfaces. (C) Complex multiple-surface: several surfaces involving baselap or truncation.

Simple multiple-surface projects involve two or more surfaces that are conformable or nonintersecting (Fig. 1.1B). Mapping these surfaces independently often, but not always, yields acceptable maps. The amount of data or geologic variability may change from surface to surface. In these cases, maps can be inconsistent and contradict the true nature of the surfaces. Even if errors do not result, ignoring conformable relationships does not take advantage of that geologic property. In many situations, use of conformability can build better surfaces, ensuring that the maps are consistent.

Complex projects involve two or more surfaces that intersect one another (Fig. 1.1C). In this instance, maps are required for several surfaces with baselapping or truncating relationships. These relationships cannot be ignored, and the map of one surface must be drawn with the other surfaces taken into account. Complex projects generally can be handled in the same way as simple projects, but with an additional step to account for the intersection.

Need to Include Interpretation

The main thrust behind this book is that two types of information are important to mapping projects. The first type is obvious—the measurements or data values. These can include elevations of stratigraphic horizons, thickness of an interval, or other variables. These are the data to be contoured.

The second type of information—geologic principles and interpretation—is rarely thought of as being data, but it is as important as the first type. Such information can include the sequence of events that led to the present geologic configuration in an area, the relationships between two intersecting horizons, the location or type of pinch-out of a sand unit, or how isolated "fault cuts" are tied together and mapped. This interpretive work, done by the project geologist, requires understanding of the local and regional geology.

Use of Mapping Tools

The starting point for a computer-mapping project consists of a set of data, geologic interpretation, and a computer program. Typical available programs have the ability to enter data and manipulate it, to create maps and other displays, and to manipulate and combine maps. This is a simple set of capabilities, and objective maps can be generated through routine use of such a program.

However, mapping geologic interpretation requires more than simple application. The methods discussed in this book do not require special capabilities, but merely make use of typical program options. Adding interpretation is thus a matter of applying basic procedures singly or in combination. The bulk of this book may be thought of as describing a box of tools. As with any tools, some will be appropriate for many applications, and others will have restricted use. However,

5

this collection of methods and approaches to problem solving gives great power to a mapping program.

Data-oriented tools modify existing data, create data, and create simple maps. Such tools include the ability to add data points or contours, to combine data that measure the same variable through different procedures, and to tie line data at survey intersections. Tools for mapping single variables or surfaces include the ability to put in trends and fault interpretations.

Tools that manipulate maps may be thought of as similar to such operations as cross-contouring. "Truncation" is an example of a tool with wide application. The concept is similar to that of an erosional surface—that is, a mapped surface is intersected or cut out by a truncating surface. It is not appropriate to map the surface where it has been removed by truncation, so blank regions should exist on the map. Any mapping method should take truncation into account.

A similar tool is "baselap," which describes another way in which two surfaces may intersect. Rather than the shallower of two surfaces being continuous as in truncation, here the shallower surface stops where it impinges on the continuous deeper horizon. Baselap and truncation operations are common to stratigraphic and structural mapping projects.

Truncation or baselap can be used to relate two intersecting surfaces, whereas two conformable surfaces can be mapped by using other tools. These simple situations can be expanded to include sets of surfaces or maps for more complex situations. Figure 1.2D shows a cross section through four surfaces. Simply creating independent maps of each surface would be incorrect, because portions of each do not exist due to erosion or nondeposition, and simple maps would be inconsistent with known geology (Fig. 1.2A).

The surfaces are created as a group by incorporating interpretation. Figure 1.2B shows the result of applying the truncation tool, intersecting horizon 1 by horizon 2. The tool for handling conformable surfaces aids in making consistent maps of horizons 3 and 4. The baselap operation laps horizon 3 onto erosional surface 2 (Fig. 1.2C), and a second application puts horizon 4 onto surfaces 2 and 3, completing the section (Fig. 1.2D). These tools are discussed in detail in chapter 6.

Baselap and truncation have wider application; they are also useful for nonstratigraphic applications. For example, when a map of oil-column thickness is drawn, fluid-contact surfaces must be intersected with the structural reservoir limits. Another example is mapping faults with low-angle fault planes; the plane is combined with structural horizons through use of these two operations.

ABOUT THE BOOK

This book is written for a "generic" computer-mapping program. A number of good programs are available, and all have the abilities to do most of the operations discussed here; some can operate with all the tools. Appendix A describes necessary program capabilities as well as some that are useful but optional.

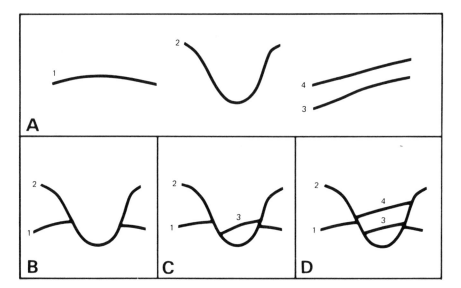

Figure 1.2 Cross section diagramatically showing horizons that are to be mapped. (A) Four intersecting horizons mapped independently; note that the maps extend into regions where the horizons do not exist. (B) Horizon 1 truncated by horizon 2. (C) Horizon 3 laps onto horizon 2; the final result is obtained when horizon 4 is lapped onto horizons 2 and 3. (D) Four horizons mapped with their correct stratigraphic relationships as observed in nature.

We assume that the reader is familiar enough with a mapping program to handle files, enter data, obtain a simple map, and so forth. The book discusses methods of applying the available program effectively, but supplements rather than substitutes for the specific program user's guide. We do not differentiate between procedures that are performed interactively and those that are performed through batch processing; the general steps described here are not dependent on how the program is operated.

The book is organized somewhat similar to the discussion in the previous section — that is, a discussion of simple mapping is followed by an examination of fundamental tools and finally by a discussion of more complex applications of tools. First, however, chapter 2 discusses basic ideas regarding contouring, stratigraphy, structure, and properties of reservoirs. The purpose of this review is to establish definitions, terminology, and concepts.

Fundamentals of computer-mapping programs are presented in chapters 3 and 4. These chapters are included for the benefit of readers new to computer mapping, to establish concepts, and for completeness. Chapter 3 covers data entry, manipulation, and calculations. Chapter 4 discusses creation of contour maps and general aspects of available computer algorithms. If a single-surface project is at hand, perhaps only these chapters will be needed.

Basic tools and applications are introduced in chapters 5 through 8. Chapter 5 discusses the handling of data with special characteristics or problems. Chapters 6 and 7 introduce concepts used in mapping sets of stratigraphic or structural surfaces; these concepts are used throughout the book. Chapter 8 discusses several ways of modeling faults and combining them into stratigraphic sets. These chapters will normally be used for multiple-surface projects.

Chapters 9, 10, and 12 include more complex applications of the fundamental tools. Chapter 9 discusses methods for introducing specific interpretations or forms into individual maps; chapter 10 presents some petroleum-oriented applications; and chapter 12 discusses the reconstruction of a set of stratigraphic and structural surfaces through geologic time. Chapter 11 discusses trend analysis.

Because of the authors' experience, the discussion in this book is oriented toward stratigraphic mapping and petroleum application. However, geologists with other interests should not be deterred. Most mapping projects involve stratigraphy or structure, and related characteristics occur in many of those that do not. Further, some tools apply to single-surface projects, so mapping individual variables can also be improved. In short, these methods are widely applicable.

Finally, a word of caution. It is easy to become so involved with computing that the "significant digits syndrome" is encountered. Do not attempt to overwork the problem beyond the resolution of the data and the reliability of the interpretation. Extreme effort can be applied to computer mapping, but this should be done only if the available information and the goals of the project warrant it.

CHAPTER 2

Contouring and
Geologic Concepts

This chapter provides a review of contouring and the principles that are important for mapping geologic surfaces; it emphasizes the concepts and terms used in this book. The chapter is divided into four parts: (1) construction of contour maps; (2) stratigraphy, specifically concordant and discordant relationships between rock layers; (3) structural modification of these layers by folding and faulting; and (4) properties of reservoirs. The geological beginner may need to refer to an appropriate text (e.g., Leet et al., 1982; Press and Siever, 1982).

CONTOURING

A first technique learned by geologists is construction of a contour map. The contour map provides a mechanism for summarizing large volumes of data about the spatial variation of a geologic surface or attribute. Lines of equal value (isolines) are used to predict the variable between observed points. Contouring involves drawing lines of equal value through a set of data points so that a realistic surface is generated between observations.

For specific variables, these lines have specific names, such as isotherms (temperature), isobars (pressure), isopachs (thickness), and so on. Although the term *contour* originally referred to isolines of elevation of structural geologic surfaces or topography, the more generic meaning — an isoline of any variable — is used in this book. Our primary concern, however, is with topographic and structural contours.

If the variable being mapped is relatively simple and many data points are available, contour maps can be constructed with little error or ambiguity. Under other conditions, contour maps are interpretive; the geologist should combine experience and an understanding of regional or local geology with observation to create the surface configuration. If data are sparse, this "contouring license" is an important element for drawing reasonable maps (Ragan, 1985).

Regardless of the amount of interpretation required or data available for contouring, a few basic techniques are always used. References include Bishop (1960), Dennison (1968), Handley (1954), Low (1977), and Russell (1955).

Contour Maps and Contouring

Four common types of geologic data are contoured: topography, structural elevation, thickness, and rock attributes. This book is concerned with the first three. Topographic maps have the most complex forms, but topographic data are relatively easy to collect, so reasonably accurate maps can be drawn at almost any scale. Maps made for buried topographic surfaces or unconformities (paleotopographic maps) should be as complex as those of present-day topography; however, they typically have smooth contours due to lack of data (e.g., Robinson, 1982).

Structure maps are contour maps of the elevation (also called subsea depth) of a distinct surface on or within a rock unit. Most depositional surfaces were originally flat lying and relatively free of irregularities. Much of the shape or structure in their contours is the result of deformation of the unit after deposition. Structural contour maps tend to be smoother than topographic or paleotopographic maps. Iglehart (1970) discusses various types of geologic surfaces and points out that they must be handled individually because their properties differ.

Two types of maps—isochore and isopach—show thickness of a stratigraphic unit. *Isochore* refers to thickness of the unit measured vertically, while *isopach* refers to stratigraphic thickness or thickness normal to layering (Bates and Jackson, 1980). In many instances, the term *isopach* has been wrongly applied to maps representing vertical thickness. For units dipping less than 20-30 degrees, vertical and stratigraphic thickness are essentially the same. However, for units with moderate to steep dip, isochore and isopach maps are significantly different, and correct identification is important. Isochore maps are used predominantly throughout this book. In addition, the unqualified term *thickness* will be used to refer to vertical thickness (isochore).

Mapped rock attributes include such variables as porosity, lithology, ore grade, coal ash, geochemical content, fossil content, and other measurable characteristics of the rock unit. Maps of geophysical field measurements are also in this category. In general, the maps of this group are extremely varied in contour characteristics and complexity and require special techniques, but they are similar in that they map rock attributes rather than rock geometries.

Harbaugh et al. (1977) summarize assumptions made in contouring: (1) The variable is continuous, at least within local regions such as fault blocks; (2) the variable is single valued — that is, only one value exists at a given location; and (3) the variable is predictable over a distance greater than the typical data spacing.

The assumption of a single-valued variable applies to most, but not all, surfaces. Structural elevation normally has only a single value at a given location, but overturned folds, thrust faults, and salt overhang at salt domes are examples of multivalued surfaces. In these situations, the data must be grouped appropriately, and separate sets of contours must be overplotted. Another example of multiple values involves an attribute such as porosity; many measurements may exist at a point, and the several values must be reduced to a single value, usually by averaging or a similar technique.

Nearby points are used to determine the spacing and orientation of contours in an area, implying that those points are in some way related — that is, autocorrelated (cf. Harbaugh et al., 1977). A variable without autocorrelation is not spatially predictable. Of course, the greater the spacing between points, the less we tend to trust our ability to draw contours that represent the surface accurately. In these situations, we tend to draw smoother contours and incorporate more geologic interpretation. Russell (1955) suggests that special lines be used to identify low confidence in contours distant from data points.

Scale and contour interval are two important considerations when making a contour map. In complex areas, the scale may need to be large to show details, while a smaller scale is appropriate to show regional trends. Recall that a large-scale map implies a map that covers a small area, and vice versa. Table 2.1 lists some common scales as ratios, feet per map inch, and meters per map inch. The purpose of the map also influences the selection of a scale.

Bishop (1960) points out that selection of a good contour interval depends on map scale, variation in values being contoured, and detail desired on the map. The purpose of the map and scale are determined first, and they dictate a range of

TABLE 2.1
Map Scales as Ratios and Units per Inch

Scale as ratio	Scale as feet/inch	Scale as meters/inch
1: 2,500	208.3	63.5
1: 4,000	333.3	101.6
1: 6,000	500	152.4
1: 10,000	833.3	254
1: 12,000	1000	304.8
1: 24,000	2000	609.6
1: 39,376	3281	1000
1: 48,000	4000	1219
1: 50,000	4167	1270
1: 63,360	5280	1609
1:100,000	8333	2540

acceptable contour intervals. The choice of contour interval usually depends on surface relief, a low relief surface allowing a finer interval than a highly variable one, as well as the amount and distribution of control.

Cross-Contouring

Cross-contouring is a procedure for performing arithmetic operations between maps. For instance, contour maps of the structural top and thickness of a unit may be available, and a structure map of the base is desired. Subtraction of the isochore from the structural top (through cross-contouring) produces a structural contour map of the unit base.

The manual process of cross-contouring for this example involves the following steps: (1) Overlay one map on the other and then place a clear overlay on top; (2) on the clear overlay, mark all locations where contours of the two maps cross; (3) at each crossing, subtract the contour values of the isochore from the structural contour's value and record the difference; and (4) after all crossings have been recorded, use the recorded differences as data to draw contours on the overlay. If a large number and even distribution of crossing points exists, the values can be contoured quickly, and there is little need for interpretation.

In addition to this example, cross-contouring can be used in a variety of ways, including:

$$\text{structural top} - \text{structural base} = \text{isochore}$$

$$\text{seismic time (sec)} \times \text{velocity (ft/sec)} = \text{depth (ft)}$$

$$\text{porosity} \times \text{thickness (ft)} = \text{porosity-feet}$$

$$\text{average ore grade} \times \text{thickness} = \text{grade-thickness}$$

Logical operations, such as retaining the minimum elevation of two structure maps to introduce truncation (cf. chap. 6), can also be done through cross-contouring.

In perhaps no other area of geologic mapping has computer technology provided a greater increase in productivity than in cross-contouring. Using a process similar in concept to manual cross-contouring, most computer programs require only a simple instruction to add or subtract two maps. The actual computing process takes only a few seconds, using the computer's internal representation of the map. The process of using the computer for cross-contouring involves grid-to-grid operations, discussed in chapter 6 and Appendix A.

STRATIGRAPHY

The sedimentary rocks of the earth's surface typically occur in layers. A single layer, or stratum, is a generally tabular body of rock possessing certain attributes

12

that distinguish it from layers above and below. Stratigraphy is concerned with the composition, sequence, age, history, environment, and correlation of stratified rocks. A stratigrapher separates rocks into layers or units on the basis of the goals of the study, commonly through use of one of these properties: *time* (chronostratigraphy): definition of units on the basis of age relationships; *rock* (lithostratigraphy): definition of units on the basis of their lithologic character; *fossils* (biostratigraphy): definition of units on the basis of their fossil content.

Typically, the goal of a geologist or stratigrapher is to combine geologic data for an area into a *stratigraphic framework* — that is, a set of horizons (identifiable tops of layers) and descriptions of the layers between those horizons. A common procedure for building a framework is to first identify horizons and describe layers in wells or outcrops and then to identify correlative horizons and layers in nearby wells or outcrops. In this way the geologist observes or interprets rock geometry, orientation, and lithologic or fossil variation or both. The framework can be used to work out the geologic history of an area, to locate depositional environments and ecologies in the geologic past, and to understand the distribution of compositional attributes of the rocks.

In order to make use of stratigraphic concepts and interpretations in mapping, the interrelationships of layers (units) must be understood. The following sections describe three common rock unit categories — chronostratigraphic, lithostratigraphic, and biostratigraphic — and the geometric relationships of the units.

Stratigraphic Units

Chronostratigraphic Unit

A chronostratigraphic unit is a body of strata that was formed during a specific interval of geologic time (Hedberg, 1976). All rocks that were formed during this time belong to the same chronostratigraphic unit. Each such unit is bounded on the top and at the base by an isochronous surface called a chronostratigraphic horizon (chronohorizon). Chronohorizons, which are bedding surfaces that separate younger from older strata, represent extremely brief periods of nondeposition or change in the depositional sequence and may be considered to represent instants in geologic time. Typical chronohorizons include layers of volcanic ash and surfaces that produce primary seismic reflections.

A period of erosion or nondeposition that is preserved in the rock record is an unconformity. An unconformity surface separates younger from older strata and represents a significant hiatus (gap in the geologic record). Both chronohorizons and unconformities separate younger from older strata, but the surfaces differ significantly in two respects: (1) A chronohorizon represents an instant in geologic time, but an unconformity can represent a long period of geologic time; and (2) rocks immediately above or below a chronohorizon are everywhere the same age, while rocks adjacent to an unconformity vary in age from place to place along the surface.

13

A group of chronostratigraphic units deposited one on top of the other in unbroken order—that is, without disturbance, erosion, or hiatus—are said to be *conformable*. A *sequence* is a group of conformable chronostratigraphic units bounded on the top and at the base by unconformities (Vail et al., 1977). The sequence is a convenient unit for stratigraphic studies because it defines a group of genetically related strata deposited during a given interval of geologic time. The cross section of Figure 2.1 shows one complete sequence (units 3 through 8) and the top and base of others, defined by unconformities.

Lithostratigraphic Unit

A lithostratigraphic unit is a rock stratum differentiated from strata above and below by lithologic character (Hedberg, 1976). The strata may be distinguished by one or more lithologic characteristics, but the unit must show a significant degree of homogeneity throughout. Each lithostratigraphic unit is bounded on the top and at the base by surfaces of lithologic change (lithohorizons). In addition to defining a boundary, a lithohorizon may also represent a thin layer of distinct lithologic character within the unit. This thin layer, commonly referred to as a "marker bed," is useful for correlation.

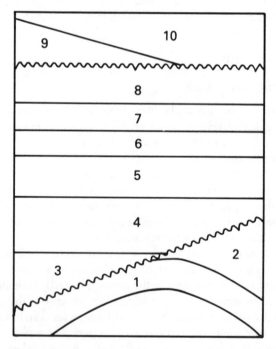

Figure 2.1 Diagrammatic cross section showing a series of chronohorizons. Units 3 through 8, bounded on the top and at the base by unconformities, form a sequence. (After Silver, 1983, p. 41)

The lithostratigraphic character of a rock is the result of the environment, the conditions of deposition, and the material being deposited. The unit will extend laterally only as far as did the environment and/or sediment type; beyond that point, the rock will be classified as a different lithostratigraphic unit. Therefore, unlike chronostratigraphic units, which are mappable worldwide, lithostratigraphic units often have limited areal extent.

Like chronostratigraphic units, two or more strata that are not interrupted by erosion or significant hiatus are said to be conformable. The same unconformity surfaces that bound the conformable chronostratigraphic units bound the conformable lithostratigraphic units. The combination of conformable lithostratigraphic units bounded above and below by unconformities is referred to as a sequence (Krumbein and Sloss, 1963).

Biostratigraphic Unit

A biostratigraphic unit is a body of strata that is differentiated from adjacent rocks by its fossil content (Hedberg, 1976). The definition of the unit may be based on a variety of criteria, including the presence of a specific fossil or assemblage, the range or abundance of one or more fossils, and fossil morphology or simply by indications of biological activity. Each biostratigraphic unit is bounded on the top and at the base by surfaces of biological change (biohorizons). As with lithohorizons, biohorizons may represent a thin layer of distinctive biologic character within the unit.

A group of biostratigraphic units stacked in unbroken sequence could be described as conformable, although they are not often referred to in this way. The same unconformity surfaces that bound chronostratigraphic and lithostratigraphic units will bound the biostratigraphic units. The combination of conformable biostratigraphic units bounded on the top and at the base by unconformities could be referred to as a sequence, although this is not commonly done.

Relationships between Units

The boundaries of chrono-, litho-, or biostratigraphic units may cross one another (Vail et al., 1977; Silver, 1983). This may not at first seem obvious, but an example of crossing lithostratigraphic and chronostratigraphic units may clarify the point. Consider a prograding beach and associated marine deposits during a relative stillstand of sea level (Fig. 2.2). At any instant, sand is deposited on the beach while fine-grained sediments are being deposited seaward. Later, continued influx of sediments and bypassing of the previously deposited material allows beach and marine facies to build seaward. As can be seen in Figure 2.2, constant-time lines (represented by the smooth lines) cross the boundary between coarse and fine sediments.

A lithohorizon marks the boundary between these two distinct sediments. When it was being formed, this boundary migrated through time, and it does not

15

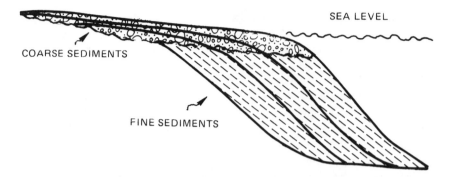

COARSE SEDIMENTS

FINE SEDIMENTS

SEA LEVEL

Figure 2.2 Lithologic units deposited in beach and marine environments. During a relative stillstand of sea level, the environments and rock types change position through time due to bypassing. The resulting lithohorizons cross chronohorizons (time lines). (After Vail et al., 1977, p. 66)

represent one moment but a range or period of time. Therefore, the lithostratigraphic units and their boundaries in Figure 2.2 transgress chronohorizons within them. Similarly, a surface that represents an instant in time can be traced through the sand unit and into the fine-grained unit, cutting across the lithostratigraphic boundary separating them.

Geometric Relations between Stratigraphic Horizons

Regardless of the type of units in a stratigraphic framework, certain geometric relationships will exist. As described earlier, the units occur in packets called sequences. A sequence consists of a group of stratigraphic units that were deposited upon an unconformity. The units within the sequence are conformable. These conformable units are bounded above by another unconformity, which forms the base of the next-higher sequence.

The conformable surfaces within a sequence generally show a high degree of parallelism. However, the thickness of the units may vary, even to the point of thinning to zero thickness (Fig. 2.3A). Such a unit is said to have pinched out (Krumbein and Sloss, 1963). Usually the pinch-out is gradual, the bounding horizons coming together at an angle of only 1 or 2 degrees. However, units such as reefs or stream channels can pinch out abruptly, forming a large angle between the bounding horizons (Fig. 2.3B).

Three geometric relationships—concordance, onlap, and downlap—are typical of the contact between surfaces within a sequence and the unconformity surface forming its base (Vail et al., 1977). Concordance occurs if the overlying conformable surfaces are parallel to the unconformity surface (Fig. 2.4A). In this case, it is difficult to identify the unconformity unless it has an angular relationship to the surfaces beneath it. Onlap is the intersection of relatively flat-lying surfaces with an

16

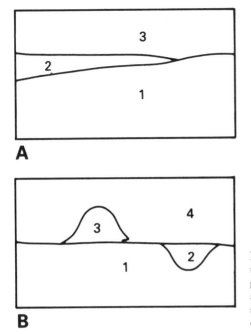

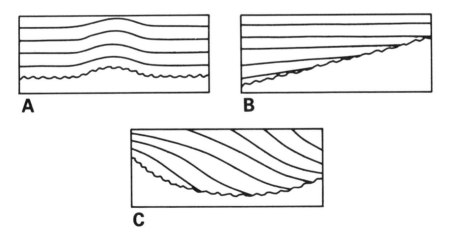

Figure 2.3 (A) Units 1, 2, and 3 belong to the same sequence. Unit 2 pinches out gradually at a low angle, bringing units 1 and 3 into contact. (B) Reefs (unit 3) and stream channels (unit 2) tend to pinch out abruptly.

Figure 2.4 Geometric relationships of horizons with the unconformity at the base of a sequence. (A) Concordance. (B) Onlap. (C) Downlap. The term *baselap* is used throughout this book to refer to both onlap and downlap. (After Vail et al., 1977, p. 58)

inclined unconformity surface (Fig. 2.4*B*). Downlap is the intersection of inclined units with either a flat or an inclined unconformity (Fig. 2.4*C*).

The terms *onlap* and *downlap* imply something about the depositional history of the stratigraphic units. At the edge of a large sea, onlap will occur on the landward side of a sequence and downlap on the seaward side (Fig. 2.5). Although it is important to recognize discordance between the unconformity and the overlying surfaces, for most mapping purposes the geometry allows both types to be handled in the same manner. The more general term *baselap* is therefore used throughout this book to describe any discordant (angular) relationship between conformable surfaces and an underlying unconformity.

Three geometric relationships—concordance, truncation, and toplap—are also typical of the contact between conformable surfaces within a sequence and the unconformity at its top (Vail et al., 1977). Concordance occurs when the underlying conformable surfaces are parallel to the unconformity surface (Fig. 2.6*A*). Erosional truncation is the termination of rock units due to erosion (Fig. 2.6*B*).

Toplap is a termination at the upper boundary of a sequence (Fig. 2.6*C*), usually resulting from sedimentary bypass with contemporaneous progradational deposition below base level. The surfaces of a sequence with toplap approach the unconformity at a high angle. Near the unconformity the surfaces abruptly take on a shallower angle and intersect the unconformity tangentially. This gradual approach often occurs over such a short distance that it cannot be detected, so it is difficult to distinguish toplap from erosional truncation. The term *truncation* is therefore used throughout this book to describe the discordant (angular) relationship between conformable surfaces and an overlying unconformity.

STRUCTURE

Structural geology, the study of deformed rocks, can be divided into three parts: the study of the shape of rock units, the study of the deformation the rocks have experienced, and the study of the forces causing the deformation. The first of these is the most important for mapping. Additional information, in the form of knowledge about forces that acted on the rock and the resulting deformation, is often useful for incorporating trends into maps. References on structural geology include Russell (1955), Billings (1972), Hobbs et al. (1976), Spencer (1977), and Roberts (1982).

Folds

Rocks can fold in three ways—flexure, shear, and flow—or in a combination of these (Billings, 1972). Each of these ways results in different geometry; some rock layers maintain their original thickness, while others thin on the limbs and thicken

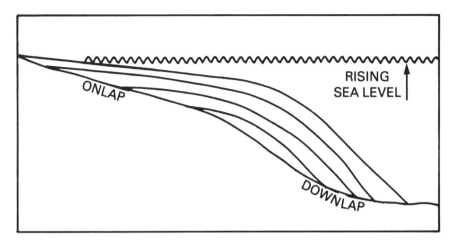

Figure 2.5 Typical position for the occurrence of onlap and downlap during coastal deposition. (After Vail et al., 1977, p. 118)

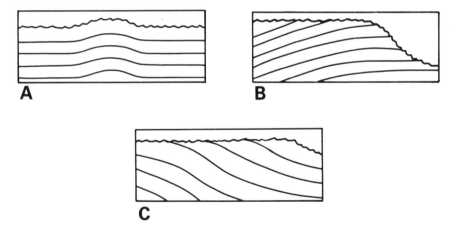

Figure 2.6 Geometric relationships of horizons with the unconformity at the top of a sequence. (A) Concordance. (B) Erosional truncation. (C) Toplap. The term *truncation* is used throughout this book to refer to both erosional truncation and toplap. (After Vail et al., 1977, p. 58)

in the crest. For mapping purposes, the type of folding or alignment of folds is usually not important, as the data used for mapping will "automatically" incorporate the fold characteristics. However, in severely folded areas with steep slopes on the limbs of folds, isochore maps will show zones of thickening and thinning. The mapping technique used in these situations may need modification to represent thickening and thinning accurately. This is sometimes done by incorporating a trend along strike or by editing the generated maps.

Some folds may be overturned—that is, tilted such that the strata on one limb are upside down (Fig. 2.7). Surfaces that have been overturned in this manner are difficult to map either manually or with the computer because the surface occurs twice at the same map location, once for the upper portion of the fold and again for the overturned portion. Contouring both occurrences, or even a third lower occurrence where it returns to its normal orientation, gives a cluttered display.

Faults

Correct mapping of faulted strata requires recognition of the fault, its orientation in space, and the displacement along it. This section is concerned with descriptions and major types of faults, displacement of rock strata, and growth faults.

Types and Descriptions of Faults

Two pieces of information are typically used to describe a fault surface: the orientation of the surface and the direction of movement along it. The fault surface

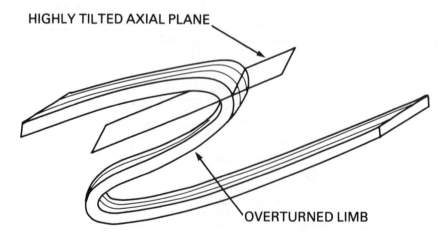

HIGHLY TILTED AXIAL PLANE

OVERTURNED LIMB

Figure 2.7 When the axial plane is tilted so that one limb is overturned, the fold is said to be overturned. Overturned folds cause the same surface to be repeated several times at the same location, creating a problem in mapping.

is often referred to as the fault plane, although it is generally irregular. The orientation is described according to its dip and strike as with bedding. For example, a fault plane might strike N45E and dip 60 degrees to the southeast.

Faults are commonly classified on the basis of the relative movement of the fault and its dip (Fig. 2.8). Three major types of faults are thrust, normal, and strike-slip (Billings, 1972). A thrust fault is one along which the hanging wall (the material above the fault plane) has moved up relative to the footwall (the rock below the plane). Two special types of thrust faults are defined by the dip of the fault plane; fault planes dipping more steeply than 45 degrees are called reverse faults, and those flatter than 10 degrees are called overthrusts (Fig. 2.8A). In a normal fault, the hanging wall has moved down relative to the footwall (Fig. 2.8B). When the fault plane dips less than 10 degrees, the normal fault is called a detachment fault.

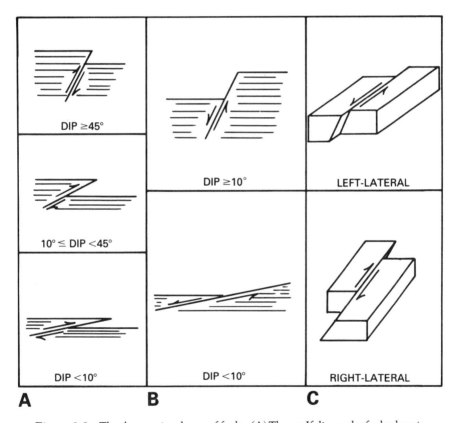

Figure 2.8 The three major classes of faults. (A) Thrust. If dip on the fault plane is greater than 45 degrees, the fault is reverse; if between 10 and 45 degrees, it is a thrust fault; if less than 10 degrees, it is an overthrust fault. (B) Normal. If dip on the fault plane is greater than 10 degrees, it is a normal fault; if less than 10 degrees, it is a detachment fault. (C) Strike-slip.

21

A strike-slip fault is one with movement parallel to the strike of the fault plane. When the movement has been to the right relative to someone looking across the fault, it is called a right-lateral or right-slip fault, and similarly to the left (Fig. 2.8C).

Displacements

Many terms are used to describe the displacement of rock bodies and the measurements made of that displacement (cf. Billings, 1972; Spencer, 1977; and Bates and Jackson, 1980). Typically, these terms refer only to measurements of displacement along the fault plane and do not apply away from the fault. However, the process of mapping deals with surfaces both at the fault and away from it.

The vertical and horizontal displacement of points on a surface away from the fault can be useful to mapping or understanding the history of fault movement, but they may be difficult to determine. Most mapped data automatically incorporate the displaced position of the surface into the maps, so no attempt need be made to understand the faulting or to restore picks to prefault positions. However, experimentation with estimates of vertical and horizontal displacement at various points on a surface and computer techniques for rapidly making maps allow the prefault configuration to be reconstructed (see discussion of restored surface method, chapter 8). Once a restored configuration is considered correct, the corresponding displacement values can be used to construct maps of the faulted surface. Although this approach is not practical in highly faulted areas, it is useful as an aid to understanding fault movement.

The sign on a displacement indicates the direction of movement. Vertical displacement is defined to be negative for downthrown fault blocks and positive for upthrown blocks (Dickinson, 1954; James, 1970). A sign convention for horizontal movement will not be suggested here, as this measurement is not used in methods described in this book.

The procedure used in this book to incorporate displacements away from the fault is based on two concepts: line of zero displacement and displaced versus nondisplaced fault blocks. The influence of a fault does not extend indefinitely, and beyond some point a given surface is not displaced. The zero-displacement line marks the boundary between zones of displacement and inactivity and may be thought of as the "hinge line" of the fault. The determination of displaced and nondisplaced blocks is based on the concept that all movement occurred on one side of the fault, with none being attributed to the other side. This approach, while not always true, is useful for computer mapping. An example demonstrates how these terms are used.

Figure 2.9A shows a vertical normal fault that dies or fades out at points Y and Z within the map area. The block to the southeast of the fault has dropped relative to the northwest block. Bends in the structural contours indicate that the surface in the southeast block has been affected by faulting; it is referred to as the displaced block. The northwest block was unaffected and will be called the nondisplaced block. The dashed line connecting points Y and Z is the zero-

22

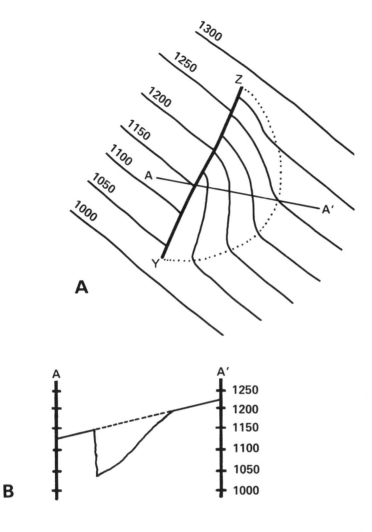

A

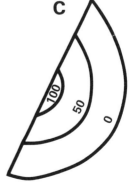

Figure 2.9 *(A)* Vertical normal fault Y,Z cuts the contoured surface. The southeast block was displaced and dropped relative to the northwest block. The curved dotted line connecting points Y and Z is the zero-displacement line. *(B)* Cross section AA′ showing present and premovement configuration. *(C)* Contours of vertical displacement generated by cross-contouring an interpretation of prefault structural contours with the faulted contours.

displacement, or hinge, line, which outlines the portion of the surface that has been moved by faulting.

Figure 2.9B is a west-to-east cross section. The dotted line represents the relative prefault position of the surface. The values of vertical displacement vary along the section: zero west of the fault, 110 feet at the fault, and zero at and east of the zero-displacement line. Figure 2.9C is a contour map of vertical displacement generated by cross-contouring an estimate of the premovement structural contours with the present contours. To simplify this example, the fault was considered to be vertical (which implies horizontal displacement values equal to zero everywhere), although in reality some horizontal displacement of the surface would occur.

In many cases it is difficult or impossible to determine a point's relative prefault position even on the fault plane (Crowell, 1959). In these cases, apparent displacements must be used. They are determined by selecting a point on one side of the fault and finding its nearest equivalent point on the opposite side.

Growth Faults

Movement on faults may take place over a short period of geologic time or during a period of time when no sediments were being deposited. If so, the "structural fault" will appear in the geologic record as if it occurred during an instant. On the other hand, movement on faults may take place over long periods of time contemporaneously with deposition of sediments. These are called depositional faults, contemporaneous faults, or growth faults. Useful references on growth faulting include Currie (1956), Bornhauser (1958), and Ocamb (1961).

The thicknesses of units on opposite sides of faults are used to distinguish between the two types of movement. Strata on either side of structural faults were originally deposited as a continuous unit and were later cut by the fault. The thicknesses of the unit on both sides of such a fault will be the same (Fig. 2.10A). Because growth faults move contemporaneously with deposition, the rock units they cut will be thicker in the downthrown block of the fault than in the upthrown block (Fig. 2.10B). Also, deep strata cut by growth faults have been experiencing continuous movement for a longer time than have shallower strata, so the displacement of corresponding points across the fault increases with depth.

Often when rock units cut by nearly vertical structural faults are mapped, a faulted structural surface is built, and then lower or higher surfaces are generated by adding or subtracting isochores. These isochores are continuous since the rock units were deposited before faulting. This approach cannot be used with a growth fault because thickness varies across the fault. Instead, each structural surface must be built independently, or, preferably, faulted isochores must be used. If displacements of surfaces due to faulting are to be used in the mapping process, then the amount of growth or displacement for each rock unit must be determined. It is

24

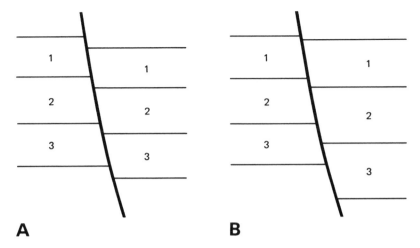

Figure 2.10 *(A)* Equal thickness of units on opposite sides of a fault indicates movement after deposition. *(B)* Variation in thickness of the unit on opposite sides of the fault indicates movement contemporaneous with deposition (growth faulting).

obvious that recognizing the presence of growth faults is critical and can significantly affect the approach used to map the faulted surface.

RESERVOIR PROPERTIES

Although the focus of this book is on mapping surfaces rather than rock attributes, such petroleum applications as volume calculations require nonstructural mapping. Fluids in the subsurface are basic to petroleum geology applications, with important related mapping problems. This section therefore briefly discusses some basic engineering and geologic properties of reservoirs. Further information can be found in Levorsen (1967), Clark (1960), Craft and Hawkins (1959), and Hunt (1979).

Space to hold fluids in rocks is of critical importance to defining a reservoir. Pores exist in sandstone and fragmental limestone, as even closely packed grains do not fill 100% of the available volume. In addition, many events can alter the volume of pore space (e.g., by dissolution of grains, mineralogical changes, or deposition of cements). Porosity is a measure of the amount of this intergranular space, defined as the percentage of pore volume relative to total rock volume. Porous rocks are potential pathways for movement of fluids; they become petroleum reservoirs when filled with oil or gas or aquifers when filled with water.

25

Nonporous rocks cannot contain substantial amounts of fluid, and so are barriers to movement of fluids.

An important geological property is the ability of fluids to move through rocks. Darcy's law (e.g., Clark, 1960; Davis and DeWiest, 1966) quantifies the rate of fluid movement through a rock; it shows that the rate varies inversely with fluid viscosity and directly with cross-sectional area, pressure gradient across the rock, and permeability, which is a measure of the ease with which a fluid can move through interconnected pore spaces.

Permeability is controlled by several variables involving the internal geometry of the rock. Sandstones that are poorly sorted, that have platelike clay-mineral grains oriented into parallel sheets, or that have undergone secondary cementation, generally tend to have lower permeability because the fluid pathways are decreased in size and are less connected, whereas clean, well-sorted sandstones have good permeability. High porosity and high permeability are often found together, although shales can have high porosity and low permeability.

Fluids—oil, gas, and water—found in the subsurface interact with each other according to physical laws. Because of density differences, oil separates from and floats on subsurface water, and gas similarly is found above oil. In addition to density contrasts, differences in hydrodynamic potential move fluids through the subsurface (cf. Hubbert, 1953, 1967; Dahlberg, 1982; Hitchon, 1984). A fluid will tend to move toward a location where it will have the lowest energy potential. Because the various fluids react differently, complex movement and patterns can result. Fluid movement is important because it accumulates petroleum into a trap.

Petroleum is said to be in a trap when it cannot move because of physical or hydrodynamic barriers. Figure 2.11 shows a typical structural trap in which sandstone is sandwiched between impermeable shale barriers and folded into an anticline. Hydrocarbons that migrate into the pore spaces will float on the water in the sandstone. Upward movement is blocked by the shale permeability barrier above, and the denser water below prevents downward movement.

Other traps are formed by structural movement. Faults associated with folds, as drag along the fault plane, can create "rollover" folds where bedding dips into the fault plane. Another structural trap associated with faults occurs when dipping beds are faulted and shale is moved in juxtaposition to porous rocks. If the barrier is updip, it can create a seal and prevent further fluid movement upward past the fault plane. Salt domes are a common location for traps; upward flowage of the salt distorts overlaying strata to form local structural highs.

The other major category of trap is stratigraphic, so called because the enclosing barriers are caused by stratigraphic or lithologic variation rather than by structural forms. Figure 2.12A shows a sandstone and enclosing barriers that have been tilted and eroded, with shale subsequently deposited over the unconformity. Any oil or gas that later migrates through the sandstone is stopped from further upward movement by the shale barriers. Figure 2.12B shows a stratigraphic trap created by a facies change from sandstone to shale.

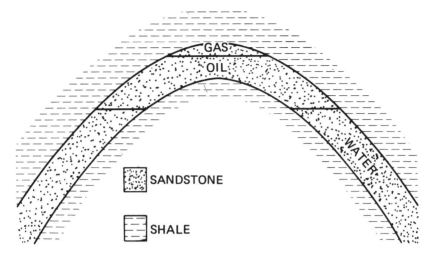

Figure 2.11 Typical structural trap, in which sandstone is sandwiched between shale barriers and folded into a dome, trapping hydrocarbons in the crest. (After Jones, 1984, p. 654)

A wide variety of combinations of trap types exist and are so varied and important that Levorsen (1967) devotes approximately 150 pages to them. Hunt (1979) discusses migration of hydrocarbons into traps. Typical maps include structural contours on the reservoir, reservoir thickness, and percentage of reservoir rock in an interval. A special use is delineation of the productive edge of the petroleum accumulation, defined by the pinch-out of a unit or the intersection of the water-hydrocarbon interface.

Figures 2.11 and 2.12 may give the impression that the interfaces between fluids are simple planes. However, the oil-water contact (OWC) and gas-oil contact (GOC) can be complex. An extreme simplification of the OWC is shown in Figure 2.13A; the fluids are in a tank and the OWC is indeed a sharp plane, defining the free-water level (FWL). However, relationships in rocks are more complex.

Even if the dominant fluid at the location is oil or gas, the fluids in a reservoir rock include some "interstitial" water between grains, leading to a zone of gradational water content. Water will rise up a thin capillary tube in response to such physical properties as surface tension, wettability of the tube to water, and capillary pressure and will rise higher in a small-diameter tube than in a larger one. The pores in a rock act similarly to the capillary tube. Water tends to move up into the rock to a level higher than the FWL expected in a simple tank, resulting in a mixture of hydrocarbon and water near the OWC.

Water saturation is the proportion of water in the fluid. Figure 2.13B shows a graph of water content through a typical oil-water contact. Water saturation is 100% at the FWL but decreases with height above the free water through a

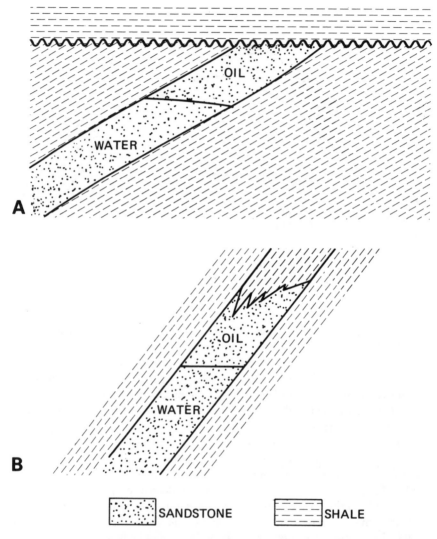

Figure 2.12 Typical stratigraphic traps, with shale barrier preventing hydrocarbon from escaping reservoir sandstone. (A) Shale deposited above unconformity. (B) Facies change to shale permeability barrier. (After Jones, 1984, p. 654)

transition zone. The width of the transition zone varies according to such factors as pore size, permeability, and relative fluid properties. In addition to water that rises above the FWL due to capillary effects, water is also found well above the FWL, approaching a limiting value with height. This limit represents water held so tightly in the pore space that it cannot be displaced; it is called the irreducible water saturation.

Even if the OWC were not of a gradational nature, it could still have a complex

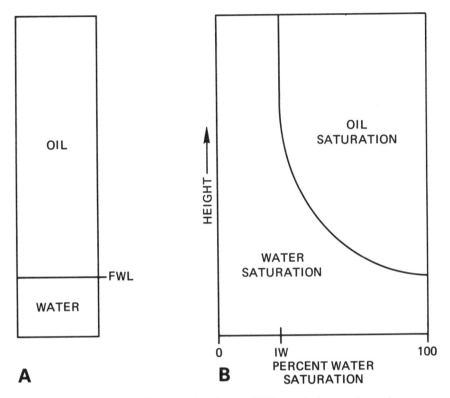

Figure 2.13 Interface between oil and water. *(A)* In a tank, the interface is sharp at the free-water level (FWL). *(B)* In rocks, the interface is gradational, with water content decreasing above the FWL. Water content cannot be less than the irreducible water saturation (IW), even well above the FWL. (After Clark and Schultz, 1956, p. B-34)

form. Lateral variations in porosity and permeability are reflected by an irregular OWC surface, as smaller pores tend to have a greater capillary effect than large pores, causing an irregular, nonplanar contact. The contact can even be tilted as a response to hydrodynamics (cf. Dahlberg, 1982).

The distributions of permeability and water saturation are important to petroleum production. Permeability controls the location of petroleum or water through its association with saturation. The location and amount of oil in a reservoir, critical for calculating in-place volumes, is dependent on water saturation and porosity.

CHAPTER 3

Input Data for Computer Mapping

A wide variety of possible measurements or observations, across many different specialties, is available in geology. Although we concentrate here on stratigraphic horizons, many other attributes are commonly mapped. For instance, petrologists determine average grain size and sorting of sediments, measure amount of cementation, determine abundances of mineral components, and so on. Geochemists, paleontologists, and geophysicists also deal with additional sets of quantitative data.

Information that is available for analysis can come from the user's own measurements at the outcrop or in the laboratory, from drilled holes, from published datasets, or from commercial sources. Varied databases have been developed for different earth science applications. Bliss and Rapport (1983), Costantino (1983), Guptill (1983), Hage (1983), and Hittelman and Metzger (1983) describe some examples.

A large amount of information, predominantly related to petroleum, also resides in commercial databases. Petroleum and mining companies also own extensive proprietary files. Petroleum-oriented databases are discussed by Forgotson (1977), Clark (1981), Iglehart (1981, 1982), Waters (1981), and Robinson (1982), among others. Such information includes digital base maps, well data, digitized contours, and correlations.

This chapter assumes that a set of data basically requiring nothing more than organization into a specific form, is available to be put into the computer for mapping. The data are then read by the computer, certain checks or calculations are made, and a special file is created. Much of this chapter is introductory and may

31

not be needed by the experienced mapper. The following sections discuss types and organization of data, input-data editing and checking, and calculations.

ORGANIZATION OF DATA

Obtaining the Data

Figure 3.1 shows a cross section through three wells, each of which penetrates two horizons. Suppose we are interested in the depth to each of the horizons, plus the average porosity in the reservoir interval between the horizons. The locations of the horizons can be interpreted from petrophysical logs (e.g., SP or gamma ray), from cores, or from drilling information. The depth at which each horizon is penetrated by the well is thus "picked," giving "horizon picks" or "correlations" or "tops" in each well. Analysis of cores or petrophysical (e.g., sonic) logs allows average porosity to be calculated.

Two types of information are available at these wells: horizon or surface picks and attributes. The most common mapping involves structural surfaces and the thickness between two surfaces. This type of information is thus geometric in nature; the bulk of this book deals with such data.

The second type of information describes rock attributes (e.g., porosity) in some region or interval. It is not geometric, but represents a property of the rock that is

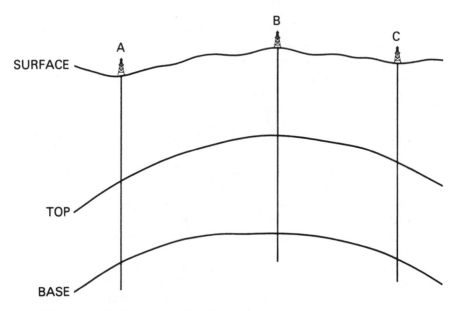

Figure 3.1 Cross section showing three vertical wells penetrating two horizons.

contained in the geometric framework. A third possible type of information is represented by measurement of a geophysical potential field, such as gravity or magnetic intensity, although this might be considered as another attribute. Other data that might be recorded include alphabetic information, abbreviations, and codes that describe other properties at the data point. For instance, at an outcrop we might record the formation or member name, type of bedding, color, and so on. This information is normally not contoured but it identifies those points to be used or edited.

Observations or measurements should be recorded in an orderly arrangement, regardless of whether or not the computer is to be used. Table 3.1 shows recorded data from the wells of Figure 3.1, plus data from additional wells, D and E, that are not on the line of the section. The table is in a common form for manual use; it is also convenient for the computer.

However, certain considerations may be needed if the computer is to be used; special recording forms and systems can be important for complex projects. Garrett (1974) and Chapman (1975) describe examples of forms for coding and recording geochemical exploration data and discuss various aspects that should be considered. Clark (1981) discusses recording information for lithofacies mapping.

Irregularly Spaced and Clustered Data

Most available information for mapping comes from outcrops or wells. We have little control over selection, as outcrops are located by nature and a well is not drilled to help a mapmaker. Thus, mapping information is not usually in a regular pattern, but is irregularly or randomly located. After we discuss this common type of data, we move on to spatially organized data recorded along lines (geophysical surveys) and data from digitized contours.

Vertical Well Data

Each row of Table 3.1 represents a vertical well from Figure 3.1. The table contains seven columns, one for each distinct piece of information. The first

TABLE 3.1
Information Recorded for Vertical Wells in Area of Figure 3.1

Well ID	X	Y	Reference elevation	Depth to top	Depth to base	Average porosity
A	76341	42684	1470	2770	2863	.173
B	77003	42698	1525	2752	2854	.201
C	77934	42701	1512	2766	2873	.160
D	78045	49569	1541	2802	2913	.215
E	74523	38638	1506	2745	2853	.161

33

column is used to identify each well. In this example the wells are given simple one-character names, but longer numeric codes, such as the 12 characters for API well number, are typically used. Well identifiers should always be included and used; they are useful for finding specific data points in lists or on base maps.

The next two columns in the table are the X and Y coordinates of each of the data points. Mapping programs usually work in terms of a rectangular coordinate system with a reference origin and two perpendicular axes (normally east-west and north-south) drawn through that point.

In some situations the coordinates may not have been provided, but data locations are posted on existing surveys or published maps. If so, the coordinates can be derived from the maps through measurement of distance between known locations and data points. In some small studies, an arbitrary user-defined coordinate system (e.g., plane table map) referenced to a designated location can be used.

Because the earth is spherical, many surveys, particularly offshore surveys, use latitude and longitude to locate points on the earth's surface. For large areas, these spherical coordinates do not work well on flat maps, so it is necessary to convert to a rectangular system; this is a substantial task because of irregularities in the earth's shape. It may also be necessary to convert from one rectangular system or projection to another. We do not discuss the well-established subjects of geodesy or coordinate transformation (cf. Richardus and Adler, 1972; Burkard, 1962), but assume needed conversions are done prior to use of the mapping program.

The wells are located on the map by their X-Y coordinates, but a third coordinate, elevation, is required to fix each well in space. Returning to Table 3.1, the column labeled "reference elevation" represents the elevation of a reference point in the well, commonly the derrick floor, Kelly bushing, or topographic surface. Drilled depths should not be mapped because the depth to a horizon could be influenced by unrelated topographic features. The depths must therefore be converted to a common datum (usually sea level) by subtracting drilled depth from the reference elevation (cf. Bishop, 1960; Moore, 1963; Dennison, 1968; Low, 1977).

This calculation converts drilled depth to elevation, sometimes called structural depth, subsea depth, or horizon elevation. Because of past confusion with depth (in which values increase downward), we prefer the term elevation (in which values correctly decrease downward) rather than subsea depth. Unless otherwise stated, data in this book are in terms of elevation.

The next two columns in Table 3.1 represent the drilled depth to each of the two horizons; these are the first of what we might think of as measured variables. These depths are measured down the hole from the reference point to each of the horizons but will be converted to elevation. The last column contains values of porosity averaged over the total interval.

Table 3.1 thus records our knowledge of these wells, and the data are then read into the computer memory, where the information is used in a form similar to that shown in the table. Each line represents a well and is made a "record" in the computer. The seven columns of information define "fields." The table is thus represented in the computer as a series of records, each containing seven fields. It is generally convenient to think of the information as being in the form of Table 3.1. The five data fields are referred to as Z-fields. Z-fields may contain measurements, data-point identification, or codes.

Deviated Well-Data

Now let us consider deviated wells – that is, holes that were not drilled vertically (Fig. 3.2). In this case, the same X-Y coordinates do not apply to each of the horizon tops or to average porosity; in fact, these locations are not the same as the location of the well at the surface. This deviation introduces a severe complication, as we know neither the location nor the true vertical depth of the observations without the use of directional survey data.

As before, the horizon picks are made in the well, but the drilled depths cannot be used directly; they are influenced by the drilling path and degree of deviation. The drill path is surveyed by measuring directions (azimuth and inclination angles) and distances at a series of arbitrary depths in the hole. These measurements are

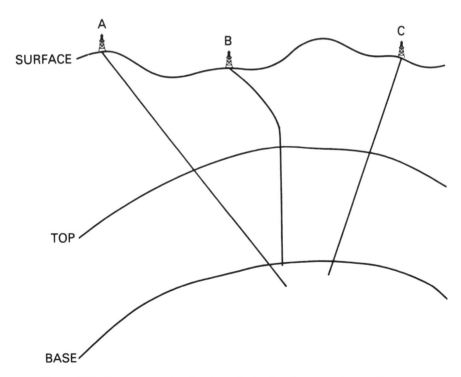

Figure 3.2 Cross section showing three deviated wells penetrating two horizons.

used to convert the points to true location (X, Y, elevation), leading to the three-dimensional coordinates of the horizon picks. Badgley (1959) provides charts to convert drilling depth to true vertical depth. This can be done with the appropriate computer program, external to the mapping program, and we assume that such calculations have been made.

Table 3.2 shows information recorded for the three wells of Figure 3.2. Because wells A and C penetrate the two horizons at different X-Y locations, each well must be represented by two records, one for each horizon. Well B is vertical in the zone of interest, so both horizon picks can be placed on a single record. This dataset does not contain a reference elevation; the reference is normally used only to convert depth to elevation, but most survey programs provide elevation directly.

If downhole thickness is recorded, a correction must be made because the interval cut by well A in Figure 3.2 ("apparent thickness") is longer than either the zone thickness ("true thickness") or vertical thickness. The data points are influenced by the drill positions rather than only by stratigraphic thickness, so the apparent thickness should be corrected. Thickness corrections also must be made for severely dipping beds (cf. Savoy and Valentine, 1961; Dennison, 1968; Pennebaker, 1972).

Average porosity in the table is arbitrarily given the coordinates of the first horizon, although the location probably more accurately belongs between horizons. Thickness corrections are not critical for average porosity, unless thickness weighting is used in calculating it. However, such a variable as thickness of net reservoir rock could be more strongly influenced by deviation of the hole than by changes in geology. Such variables should be corrected for this excess geometric thickness by use of the directional survey data.

This discussion described data obtained from wells; other data sources are possible. We can collect samples from outcrop and measure several attributes on each sample (e.g., average grain size, sorting, heavy-mineral content, bed thickness, and so forth in Table 3.3). The Z-fields include sample identification, codes to indicate the bed/member sampled, ground elevation, and the measured attributes. The non-numeric bed codes cannot be contoured. However, they can be posted on base maps or used for data manipulation or calculations (see below).

TABLE 3.2
Information Recorded for Deviated Wells in Area of Figure 3.2

Well ID	X	Y	Elevation of top	Elevation of base	Average porosity
A1	76341	42684	−2306	−	.173
A2	76613	42698	−	−2410	−
B	77934	42701	−2134	−2223	.206
C1	78445	42720	−2375	−	.173
C2	78689	42742	−	−2483	−

Specially Organized Data

The previous discussion assumes that the wells or data locations are randomly or irregularly located, and nearly all datasets are of this type. However, two forms of data with regular spatial arrangements — linear and digitized — are also common in industry.

Linear Data

Seismic surveys are usually taken along crossing sets of parallel lines. Sound waves are sent down at each of a number of locations ("shot points") along the lines, reflect off layer boundaries, and return to the surface. The elapsed travel time is recorded, and extensive computer processing converts the digital information into a seismic section. This is a cross section along the seismic line, but travel time of the sound waves is on the vertical axis. The seismic method is discussed by Dobrin (1960), Grant and West (1965), Hollister and Davis (1977), and Waters (1981).

Interpretation of the section gives what looks like a geologic cross section, as the stratigraphic and structural horizons are delineated according to geologic principles (cf. Vail et al., 1977). The user must then transfer interpretation from the seismic sections to the computer by measuring seismic travel time to the various horizons. These measurements are taken at selected points along the lines and are closely spaced relative to the distance between the lines.

The sections, and hence the measurements, are in terms of seismic travel time, which must be converted to depth or elevation. Each layer of rock has an associated sonic velocity, so multiplying travel times by the appropriate velocities gives depth. While this calculation appears simple, obtaining and using velocities may not be (cf. Hubral and Krey, 1980); computer programs are normally used for these conversions.

The form of seismic data is similar to that of wells, as several horizons are "picked" or interpreted on the seismic sections. A record for a given sample location would include the X-Y coordinates, line and shot-point identifier, and measurements (time or depth to several horizons).

TABLE 3.3
Information Recorded for Outcrop Study

Sample ID	X	Y	Bed	Surface elevation	Grain size	Sorting	Heavy mineral	Bed thickness
A1	4131	29310	K	3119	1.5	1.1	15	3.1
A2	4150	29305	K	3133	1.3	0.9	12	2.9
A3	4202	28915	L	3151	2.1	1.4	5	0.8
B1	5503	30251	K	2976	1.6	1.1	10	3.2
B2	5519	30295	J	2990	2.5	1.5	3	5.6

Aeromagnetic surveys are normally laid out in parallel flight lines. As the magnetometer moves along the line, it automatically provides closely spaced observations along each line. Each data-point record usually contains the X and Y coordinates of the measurement location, a line and sample identifier, and the measured magnetic intensity.

The measurement procedure may be automatic in an aeromagnetic survey, resulting in samples too closely spaced for the mapping application. If so, we need to delete some of the data points, and these edits can be done more easily and effectively if the points are in lines. Linear data are particularly useful if the data include a "noise" component because a moving average can be used to smooth points slightly. Such calculations are more difficult if the data are not grouped by line.

Digitized Data

Digitized contours make up another type of regular information. Here we use a map that has already been contoured by hand — that is, it is an interpretation based on randomly spaced or linear data. A machine called a digitizer is used to convert the contours to a series of X-Y coordinates. This is done as the operator moves a stylus over the lines. As the contours are traced, coordinates of points along the lines are sent to the computer.

In this case we have a series of records containing X and Y, plus one Z-field. All points along a contour are assigned the same contour value. The complete set of (X,Y,Z) points is then used to create a contour map. This procedure thus "puts a map into the computer."

The use of digitized data is not as straightforward as it might seem. The spacing of the digitized points can influence the resulting map (i.e., points should not be generated too densely), as can the selection of those contours to be traced. In addition, additional points or contours are sometimes needed to replicate the original map. Digitized maps are discussed in chapter 9.

DATA EDIT AND ERROR CHECKING

Table 3.4 shows wells, F and G, added to those in Table 3.1. Well F has no entry for the basal horizon, probably as a result of drilling that did not reach that depth. Similarly, no entry exists for average porosity. Looking back at Table 3.2, we note that fields were left blank in the deviated wells. Other reasons for missing data might be difficulty in taking measurements, time constraints, or lost data. In some instances, we purposely introduce "missing" data (see chaps. 5 and 6).

The identification and special treatment of missing entries is an important consideration in mapping. Many programs, if not instructed otherwise, read the blank field normally and interpret the blank as a zero value. If these zero values are

TABLE 3.4
Information Recorded for Vertical Wells in Area of Figure 3.1, with Missing Data

Well ID	X	Y	Reference elevation	Depth to top	Depth to base	Average porosity
A	76341	42684	1470	2770	2863	.173
B	77003	42698	1525	2752	2854	.201
C	77934	42701	1512	2766	2873	.160
D	78045	49569	1541	2802	2913	.215
E	74523	38638	1506	2745	2853	.161
F	77471	46829	1516	2749	–	–
G	73628	45631	1573	2789	2874	.184

used, the resulting maps will be totally unsuitable. Because we do not wish zero values to be mixed with actual measurements, it is necessary to identify such missing values.

A Null value is a special number that cannot ever appear in real data (e.g., -10^{30}). Most modern mapping programs include Nulls, although they might be called Nil, Znil, Znul, or some similar term. The program should be instructed to look for missing or blank fields and to insert Null values in their places to indicate missing observations in the dataset. All other options in the program should be designed to exclude Nulls from later calculations.

Some programs can detect blanks automatically, whereas others must be instructed to check on a field-by-field basis during data entry. Still other programs must detect blank fields indirectly. If the blank field is read as zero, and zero is never valid in the field (e.g., depth), then calculations (see below) can be used to change all zero values to Null. If zero might be a valid entry, then the user must enter some code value (say, -9999) in the field and use calculations to detect and change it to Null. Appendix A discusses necessary program capabilities regarding Nulls, calculations, and identification of blank fields.

Checking for missing data is only one part of data preparation. It is usually necessary to check and edit errors, which can range from simple transposition of digits to geological miscorrelation of a horizon. The process of cleaning up the data is usually tedious, but it is important to the success of any mapping study; even one bad data point can have an adverse impact. Although data checking is discussed only briefly in this book, we stress that it cannot be ignored.

Simple statistics can be a useful tool for filtering erroneous data values. The minimum and maximum values in each Z-field should be obtained and analyzed. If these limits are reasonable, we expect no problems, but unusual values may be suspect. For instance, a porosity value of 70% or a structural depth of 35,000 feet should be investigated.

In order to look at more than just the extreme values, a histogram of Z-values for each field is useful because it shows relationships for all values. In addition, the histogram can indicate if two populations exist in the data — that is, if it is bimodal. Bimodality can occur, for instance, when exploration assays find a geochemical anomaly superimposed on background (Rose, 1972).

Data from a stratigraphic framework can be checked for consistency. If bedding has not been overturned or repeated, knowledge of which horizons are youngest (shallowest) and which are oldest (deepest) allows a simple check to determine that the depths are recorded in stratigraphic sequence — that is, the depth to a given horizon must be greater (deeper) than that recorded for a younger horizon.

We recommend plotting simple work maps. They show the data distribution, and posted Z-values often indicate trends. A value that is significantly different from those around it could be erroneous and should be checked for correctness.

A special type of work map, drawn on the printer, can also be used for checking. Here a map is scaled on the page, and a one-character symbol is printed at each data location. If the range of potential Z-values is divided into a few (say, ten) intervals, a unique symbol can be assigned to each interval. Printing this map and symbols gives a rough indication of variability and highlights outlier values, even though the symbols occur only at the scattered data locations.

The computer contour map is an excellent tool for detecting erroneous points. If a Z-field contains a bad value, the contour map made from that field may indicate the error. Contours will cluster around incorrect data points, creating bulls-eyes that highlight the erroneous values. Unless the variable under consideration is "noisy," outliers will generally call attention to themselves.

Although the previous discussion refers to Z-values, it is also possible that an X or a Y coordinate may be in error. Again, obvious errors can be detected by looking at minimum and maximum values in the fields. Errors in the coordinates can also be detected by contour maps. If a bulls-eye appears in maps for all Z-fields, the data point location or reference elevation, rather than all Z-values, should be suspected.

In addition to detecting errors, statistical analysis can be valuable for understanding the variable (e.g., McCammon, 1974). For instance, average elevations describe the general location of the horizon, and the standard deviation or range of values gives an indication of the diversity of values, which is helpful in selecting a contour interval. We recommend careful study of a data set, both for error detection and for geological understanding.

DATA CALCULATIONS

After the data have been stored in the computer, ideally without errors in any values, we may perform calculations, commonly to change scales or units, combine variables, create new variables, and edit or filter data. In other words, we modify, delete, or create information before mapping.

Consider the information in Table 3.4. Measured depths to the two horizons were recorded and input to the computer, but we must convert depth to elevation for mapping. Measured depth is converted to elevation by subtracting depth from the reference elevation. In Table 3.4, suppose reference elevation is considered as Z-field 2 (Z2) and the two depths are Z3 and Z4. At each well, the elevation of the top horizon is thus calculated by $Z2 - Z3$, and the elevation of the deeper horizon is $Z2 - Z4$.

The program used for these calculations must be instructed as to what calculation is to be made (here, subtraction), which Z-fields are involved (for instance, Z2 and Z3), and where the result is to be placed. In most cases we have no further interest in the measured depth, so the resulting elevation would replace depth in field Z3. Depth in field Z4 would similarly be replaced by its corresponding elevation. The calculation cannot be made in any fields containing a Null value, so in this case a Null should result.

Converting depth to elevation is basically a change of units. However, another calculation can derive a new variable. Thickness of the interval bounded by the horizons is commonly mapped. This thickness can be obtained simply by subtraction of the elevations of the horizons, $Z3 - Z4$. We normally retain the elevations, so the program must create a new field, Z6, for the thickness values.

Other typical arithmetic calculations include adding, multiplying, or dividing values in one field by another or by a constant. Logarithmic, exponential, or trigonometric functions may also be useful.

Another type of calculation involves data manipulation or edits. Turning to Table 3.3, we see that field Z2 indicates the bed from which the sample measurement was taken. Suppose we wish to map grain size in bed K. If grain-size measurements corresponding to non-K beds are converted to Null values, then a map made from that field will include information only from bed K. Rather than using Nulls, the operation might just delete the non-K records.

Other calculations might convert a value to Null if it exceeds a given cutoff value or perhaps if a corresponding value in a second Z-field exceeds a cutoff, with similar considerations for values less than a minimum cutoff or equal to a constant. For instance, we might test porosity versus 8%. Other tests take certain actions on a field for records in which it or another field contains a Null value.

So far we have discussed calculations on Z-fields. However, we need to operate on the X-Y coordinates as well. For instance, division by 3.28 converts feet to meters. Translation of the coordinate system is useful if the numbers contain many digits, as rounding errors can influence the mapping; subtraction of appropriate constants (say, the coordinates of the lower-left corner) can give more manageable values.

Rotation of the coordinate system is another useful mapping technique. Figure 3.3A shows locations of data points on a map. The northeast-southwest trend of the data leads to large empty areas on the map. Rotation of the coordinate system gives a map with more efficient use of space (Fig. 3.3B). The X-Y coordinates can

41

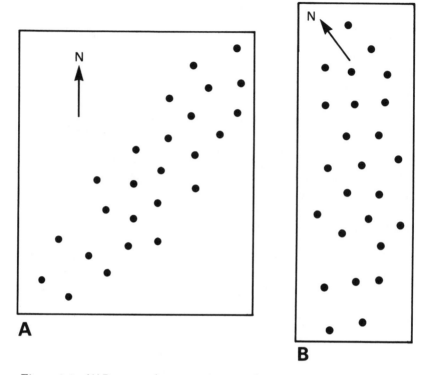

Figure 3.3 *(A)* Base map showing empty areas due to trend of data locations. *(B)* Rotation of coordinates reduces empty areas.

also be used to delete data outside the current area of interest, thereby reducing the amount of information to map.

The ability to make calculations on the input data is an important first step in creating contour maps. This can involve specification of the data to be used, creation of fields, or modification of original information. Be sure to take advantage of this powerful tool, whether in the mapping program or in an external program.

CHAPTER 4

Simple Grids and Contour Maps

Chapter 3 describes the organization of data into a form accessible for use by the computer and its input to the machine. This chapter discusses creation of a grid and the use of the grid for drawing a contour map or other displays. Chapters 3 and 4 thus parallel the steps followed in hand-contouring: data collection and organization, areal definition, and drawing contours.

Chapters 3 and 4 relate to projects involving only a single surface or variable. However, chapter 5 discusses problems that may arise with specific types of data, chapter 8 is needed for handling faults or discontinuities, and chapter 9 discusses the inclusion of special interpretations. The more common case of working with several horizons is discussed in chapters 6 and 7.

This chapter briefly describes some of the many methods and algorithms used to generate contour maps with the computer. It is not meant to be a source from which one could write a program, but rather it is meant to provide an overview for the geologist who wishes to know about the "processes" that can affect his computer contours. This discussion may also provide some insight into the selection of program controls. Many choices are available for creating maps and no one method can be automatically judged best, although each has properties that are of value under given circumstances. Before getting into computer aspects of contouring, however, let us consider hand mapping.

Hand-contouring typically starts at a local high or low, in a simple area, or at a region with dense data. Points surrounding this location define gradients, and contours are drawn to suit the data and gradients. After a number of contours have been drawn and shapes begin to appear, geological reasoning and regional trends can be applied to extend contours into adjacent regions — normally with liberal use of an eraser — until the complete map is drawn.

Such a task is complex, and the ability of the geologist's eye and mind to recognize many points and relationships simultaneously is extremely important. The computer does not approach the task in quite this way because it cannot integrate relationships as does the human brain, but it is able to consider many combinations of groups of points and relationships.

Even in hand-contouring, an irregular data distribution is usually more difficult to contour than an evenly spaced, dense distribution. If the data had been observed in a regular square pattern, gradients and trends would be easier to detect and project than with scattered data, so the ordered dataset would be easier to use. This concept of ordered information leads to the grid, a tool used by most current contouring programs because of efficiency of storage and ease of manipulation.

DEFINING A GRID

A grid is a set of values or numbers that is regularly arranged, commonly in a square or rectangular pattern, although other forms are also used. The locations of the values represent geographic locations in the area to be contoured, and the values represent the variable being mapped. A typical grid might consist of a set of elevations, corresponding to the elevation of a geologic surface.

The use of a grid in contouring involves four steps: (1) delineation of the area and variable to be mapped; (2) definition of a framework or grid pattern over the area; (3) calculation of values to be assigned to each grid point; and (4) use of these values to draw contours. Steps 3 and 4 are presented in detail in the following two sections. Figure 4.1A shows a base map, which outlines the area to be mapped; the posted values represent the variable. This corresponds to the first step.

The second step is definition of the grid framework. Figure 4.1B shows a grid pattern made up of rows of horizontal lines and columns of vertical lines. The rows and columns cross, defining grid intersections or grid nodes. Specifying the grid limits (minimum and maximum X and Y coordinates) and the grid spacing defines the grid framework.

The grid in Figure 4.1B contains intersections that are spaced the same distance apart in both X and Y directions — that is, the nodes define a series of grid-squares. The grid could also be a series of rectangles if the row spacing is different from the column spacing. In this book, the term *grid-square* has the same meaning as "grid-rectangle." Square or rectangular grids are most common in mapping programs, and unless otherwise stated, are assumed here.

Other patterns can be defined, and the procedures discussed in this book are also generally applicable. Grids made up of triangles, rather than squares or rectangles, are becoming more common. Some programs do not use formal grids, but have grid-equivalents. These and triangular grids are discussed below.

Whenever two or more surfaces are gridded, it is necessary to identify the grids

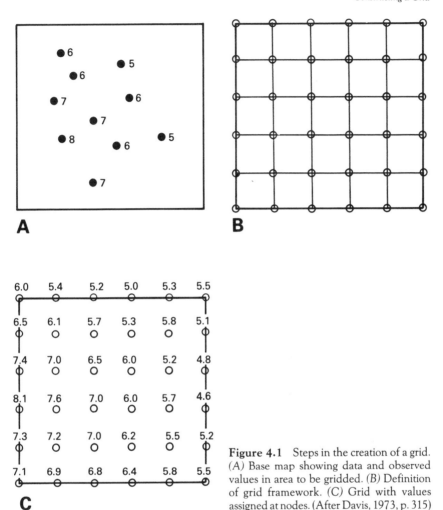

Figure 4.1 Steps in the creation of a grid. (A) Base map showing data and observed values in area to be gridded. (B) Definition of grid framework. (C) Grid with values assigned at nodes. (After Davis, 1973, p. 315)

individually. This can be done with an identifying code name or number assigned to each grid or by specifying a location or member in a file containing grids. The particular method used is not important as long as it is convenient. Throughout this book we assume that code names (e.g., A, UNCF, OWC, FE2O3) or numbers are used. Grids and grid handling options are discussed in Appendix A.

CONSTRUCTING A GRID

The framework is only part of the grid. The third step in making a grid is assignment of a number to each of the grid nodes or intersections (Fig. 4.1C).

45

These numbers represent values of the variable under consideration (e.g., elevation of a structural surface, thickness of a sand body, ppm of gold) and are calculated from the data in such a way as to represent the surface or variable at each of the node locations. As far as the computer is concerned, all information is contained in the grid, and the original observation points are unnecessary.

Types of Algorithms

Early mapping programs were consistent and objective, but did not allow much flexibility. Demands by geologists for ways to control how data are used has led to many different algorithms for assigning values to grid nodes. Several basic methods (with variations) exist, and discussions are given by Walters (1969), Crain (1970), Sampson (1975), Yoeli (1975), Harbaugh et al. (1977), Waters (1981), Bugry (1981), Ripley (1981), and Davis (1986), among others. The primary concern of this book is not with gridding algorithms, so our discussion is brief, giving only flavor and background; it should, however, aid in obtaining an acceptable geologic interpretation and solving data problems.

Independent Node Calculations

One approach makes a separate calculation at each grid node. Aside from the fact that common data points are used, the calculation at one node is independent of the value at an adjacent node. At each intersection the computer selects data points to be used, makes a calculation with these points, and assigns the value to the grid node.

Selection of data points should include two important considerations.

1. The points should have an even geographic distribution around the grid node. If data are clustered, it is possible that the points nearest the node all lie together relative to the node. For example, consider the seismic-line data shown in Figure 4.2A. Suppose we wish to use the nearest eight points to calculate the nodal value. The figure shows that all eight points come from the same line, and east-west gradients are not used.

A good method for ensuring better data distribution is through the use of search sectors. Figure 4.2B shows eight sectors centered on the grid node. Here the nearest point in each sector is used to calculate the nodal value, so the eight selected points define gradients from various directions. A common extension of

Figure 4.2 Map of line data (small open and solid circles) and grid node to be calculated (large circle) with eight values. (A) Nearest eight points are all from same line. (B) Use of eight sectors improves distribution of points to be used. (After Harbaugh et al., 1977, p. 111)

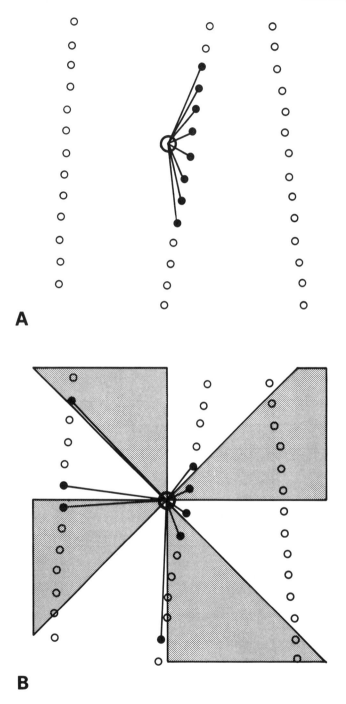

A

B

this method selects several points in each sector. This method is also important for use with randomly scattered points.

2. Only data points near the grid node should be used. An observation may be consistent with nearby values, but it is not reasonable to expect it to influence the opposite corner of the map. Accordingly, algorithms commonly include distance limits, beyond which a data point will not be used. If a maximum number of data points (either total or by search sector) is specified, this maximum may be attained before the distance limit is reached.

Of course, additional criteria may be used to select the points, including the use of minimum or maximum limits on the values being mapped. Points containing Null values (see chap. 3) should not be used.

Once the points to be used for calculating a given grid intersection are selected, the calculation may be done in several ways, of which we describe three general approaches.

1. The simplest method of calculating the node value is to use a weighted average of the selected data values. Common weights are based on inverse distance or inverse distance-squared (e.g., Davis, 1973), although more complex functions are used (cf. Harbaugh et al., 1977). Ripley (1981) points out that the weights should decrease with distance at least as fast as inverse distance-squared to prevent "spikes" at the data points.

The advantages of the weighted averaging method include speed of computation and simplicity of programming, although search sectors are normally required. However, averaging also has disadvantages. First, even simple forms may not be mapped correctly. For instance, if several co-planar points are gridded through averages, the map in that region may not show a plane. A second disadvantage is that when an average is calculated, the result cannot exceed the largest value used or be less than the smallest value. This prevents projections outside the range of Z-values; while this is sometimes reasonable, situations exist where it is not. For example, in mapping isochores the grid cannot project to node values equaling zero because the average trends toward a local mean thickness. Zero-lines can be a problem with some unsophisticated programs.

2. A somewhat more complex way to calculate a value involves the fitting of simple, low-order functions, commonly planes or quadratics. This consists of fitting the function to the selected points and then interpolating a value from the function at the location of the node. With this method, trends in the data have greater influence than with weighted averages because the function can project slopes. Slopes (dip and strike) estimated at each data point can also influence the fit (cf. Harbaugh et al., 1977).

3. A geostatistical procedure called kriging may be used (cf. David, 1977; Journel and Huijbregts, 1978; Clark, 1979). This method takes into account inherent variability in the data being mapped. This variation is estimated through calculation of a semi-variogram, a statistical tool that relates variation to distance.

Qualitatively, we expect nearby points to be more similar than distant points, and the semi-variogram quantifies this expectation.

Kriging is a complex weighted average that takes into account spatial relationships between data points and from them to the grid node. Kriging has both advantages and disadvantages. All spatial information in the data is used in mapping. In addition, the value at each grid intersection has minimum error variance. A grid of these error variances can be generated so that maps show regions of high or low confidence.

On the negative side, a major assumption in kriging is that the surface or variable is stationary — that is, it does not have regions with broad trends. A surface that dips uniformly over the entire map area could give poor results. The lack of ability to project values is also a disadvantage. In addition, maps from kriging can be unacceptable visually because they typically have highly variable contours with excessive local variation; Davis and Culhane (1984) and Davis and Grivet (1984) present methods to improve mapped appearance. Finally, under certain conditions the map may not honor data values.

Node Calculations Spreading from Centers

A second approach to gridding uses previously calculated node values to calculate new values. In this method, calculations spread outward from each data point, filling in nodal values as multiple passes are made over the grid. Values are based on other grid nodes, so they are not independent. This general approach is described by Walters (1969) and Crain (1970), but the two-step process is summarized briefly here.

The program first calculates values for the four grid intersections surrounding each data point. This is done through the previously described procedures of selecting nearby data points and calculating values at the intersections. As discussed above, search sectors and distance limits are appropriate. Surface fits are preferred to averages in order to retain trends or gradients.

After the nodes around all wells are calculated, the wells are removed from further consideration. The program then makes a series of passes over the grid. At each pass it calculates values for any grid nodes that have not yet been assigned a value and that are adjacent to an assigned node. In other words, each iteration enlarges the calculated region around the original well locations. The procedure used to calculate a value is similar to the one given above, except that instead of selecting nearby data points, nearby nodes with calculated values are used to generate the new intersection values.

This procedure works well with surfaces that are not extremely complex, and it generally honors data with little difficulty. It also has the advantage of allowing trends to be projected, although caution should be used because extrapolations can become extreme more rapidly than with the independent calculations described above.

Other Gridding Methods

These two general approaches are probably most often used for creating a grid. However, other methods are available. A procedure that is widely used in the mining industry for gridding attribute data is known as the nearest-neighbor or polygonal method. This procedure extends the area of influence of each data value halfway to the nearest data point. Boundaries are thus defined midway between control points, leading to a series of polygons, each with a single data point at the center.

An alternative definition of the nearest-neighbor method is more efficient for the computer. Assign to a given node that value associated with the nearest data point—that is, its nearest neighbor. Several nodes near a data point are thus assigned the same value, with an abrupt change in value when moving to nodes near another point. The result of either method is a grid that is made up of a series of flat plateaus, each at the height corresponding to its included data point.

The polygonal method is simple to program and operates faster than other gridding algorithms. In addition, if data are qualitative (e.g., rock type), it is optimal for mapping in the absence of other information (Switzer, 1965). However, this method is not appropriate for most quantitative applications because the discontinuous nature of the grid does not reflect the continuity of most geologic variables. In addition, this method ignores geologic trends and allows the geometry of the data points to dictate the range of influence of the samples.

Spline-fitting is a commonly used quantitative method (cf. Ahlberg et al., 1967). In one dimension, a cubic spline is a polynomial function that will pass through the observed points. Between the points, the function will have a form as if it were a thin, flexible rod. This spline function has good mathematical properties and is widely used for interpolation. The bicubic spline is a two-dimensional version, the surface representing the form of a thin sheet.

Two-dimensional fitting of splines was first devised for gridded (regularly spaced) data sets (e.g., De Boor, 1962; Bhattacharyya, 1969) and thus was not useful for most geological data. Hessing et al. (1972) extended the method to data in the form of a distorted grid so that a square grid could be derived. Briggs (1974) solves differential equations iteratively to calculate splines for nonregular data. Gonzalez-Casanova and Alvarez (1985) discuss splines from several points of view. Dubrule (1983, 1984) and Watson (1984) point out that spline interpolation may be regarded as a special case of kriging.

Mathematical functions that are defined over the entire map area can also be used. A value can be interpolated from the function at the location of each grid node, and these values can be used to create a grid. Note that this involves a single fit over the entire grid area, whereas above we described piecewise, localized surface fits. Cole (1969) discussed a combination of trend surfaces (see chap. 11) and numerical procedures. Other mathematical functions and numerical procedures (cf. Crain, 1970; Junkins et al., 1973; Pfaltz, 1975) can also be used to create a grid.

Special algorithms have been devised for handling specific types of data. Measurements are commonly taken along lines (e.g., seismic-line or aeromagnetic data; see chap. 3), which can present problems in gridding. However, the special properties of line data can be used to advantage (e.g., Bhattacharyya, 1969; Rasmussen and Sharma, 1979). Digitized contour lines present a similar type of information (see chaps. 3 and 9). Again, special algorithms can improve gridding with this type of data, but use of sector searches and careful selection of lines to be digitized can give good results.

Grid Refinement

Some gridding algorithms, notably those with independent calculations from node to node, create a grid that will not lead to smooth contours. The contours might show sharp, unjustified small-scale bends. Therefore, after a grid has been generated, smoothing or filtering is sometimes used. In this case, creation of the grid is a two-step process; one of the methods described previously is used to make an initial grid, and then the grid is refined to improve its properties.

Briggs (1974) presents methods that operate on a grid and modify it locally in order to reduce curvature. In so doing, small irregularities are removed, but the major features are retained. This filter is based on the use of bicubic splines. Other possible filters include weighted moving averages. However, these filters are designed for removal of extremes (local highs or lows), rather than for smoothing irregular contours. Such grid modification can reduce the ability of the grid to honor the data.

Another refinement forces the grid to honor the data points. When contours are drawn based on the node values, it is possible that they do not agree with the posted data. For instance, if a grid-square contains several data points with different values, it may be impossible for the few local grid nodes to force contours to have the correct relation with all values. Pragmatic or theoretical (e.g., splines) procedures using the original data can adjust the grid in the immediate vicinity of a data point to better honor observation, although tight clusters of data cause difficulty for any procedure.

Important Parameters

Regardless of the mapping program available, certain parameters or controls must specify how the algorithm is to be used. They include grid interval, search-distance limits, number of search sectors, and maximum or minimum number

51

of points used in calculating a node value. Of course, choice of algorithm is also influential.

The grid interval controls the detail that can be retained in the grid. No features smaller than this interval are retained, and accurate definition of a feature usually requires that it cover two to three grid intervals; the interval should thus be made small enough to show required detail. On the other hand, creating a grid with too fine an interval wastes computer time and memory and in some cases (e.g., spreading algorithms) causes unacceptable extrapolations. The following discussion assumes a square grid, but the same considerations apply for the two directions of a rectangular grid.

A rule of thumb says that the grid interval should be chosen so that a given grid square contains no more than one data point. Of course, twin wells, replicate samples, and the like can provide closely spaced information, but generally most data locations are separated. A useful procedure in this case is to estimate (by eye) the typical spacing between near-neighbor data points and then use as the interval a convenient even value (e.g., 100 rather than 98.4).

Figure 4.3 shows a contour map of the unconformity data (field UNCF in Table B.4) of Appendix B, with well and seismic picks merged (chap. 5). The grid has an interval of 750, giving 41 rows and 29 columns. The map shows detail in the data but is not artificially complex. Some might find the small closures objectionable, although they reflect the data and therefore result from most gridding algorithms. Note that this map shows a strong trend or grain oriented approximately N30E. Enhancement of this trend (see chap. 9) would improve the map.

Figure 4.4 shows a contour map of the same data, but based on a too-coarse grid interval (3000), with 11 rows and 8 columns in the grid. Here the map is too smooth and does not represent variations found in the data. In fact, many data points are not honored. Rhind (1971) shows the effect of grid interval with extensive comparisons using two mapping programs. He presents maps ranging from grids containing only five rows and columns to grids with 100 rows and columns. The coarse grids are excessively smooth, but detail increases to a limit controlled by data distribution and size of features as the grid interval decreases.

If the data are linear — for example, from a seismic-line or aeromagnetic survey — make certain that the selected interval is appropriate for the detail and need of the project because linear data are sometimes collected too densely. Excessive data points can create problems in making a good grid as well as using excessive computer time. Deleting some points after application of a moving average on the line, or picking peak-valley points, might be appropriate.

With digitized data, the grid interval should be based on the needed detail. However, it is sometimes selected on the basis of contour spacing. As above, data spaced too closely can cause difficulties; remove points from overly dense digitized contours. Chapter 9 discusses digitizing contour lines.

In addition to the grid interval used, density of the input data influences the

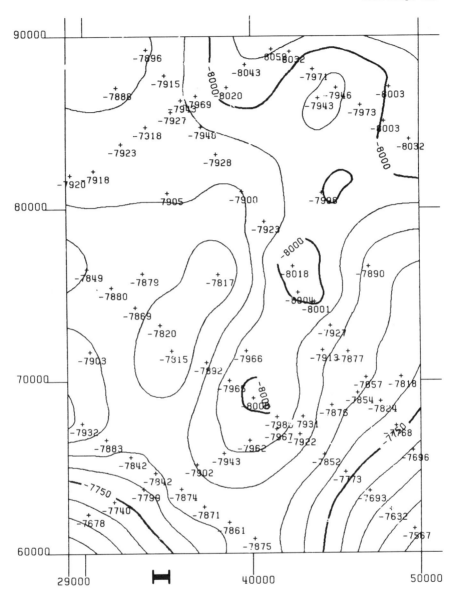

Figure 4.3 Contour map gridded with appropriate interval of 750 (indicated by bar); data from UNCF field of Table B.4.

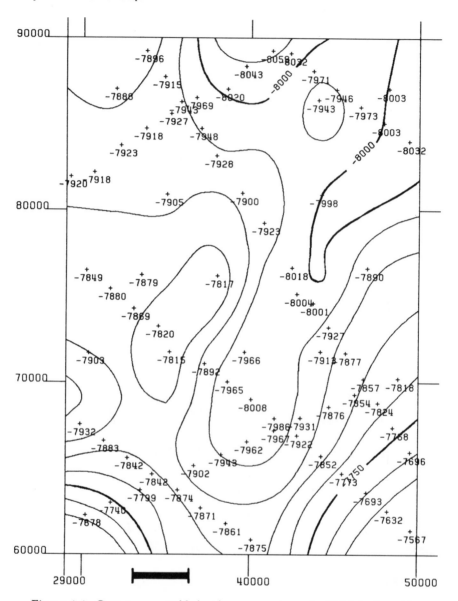

Figure 4.4 Contour map gridded with too-coarse interval of 3000 (indicated by bar); data from UNCF field of Table B.4.

form of the grid and subsequent contour map. Widely spaced data cannot give maps containing small details, regardless of the grid interval used. Dahlberg (1975) and Robinson (1982) studied the effects of different data density on contour mapping and found that details such as drainage patterns are lost through lack of control; even large features and trends may disappear or be distorted.

Most current algorithms require that good data-selection routines be incorporated, normally implying the use of search sectors. The number of such sectors usually ranges from 4 to 16, but 8 is a common number; better results generally come with more sectors, particularly with nonrandomly distributed data. Along the same line, use of more than one point per sector usually increases stability.

The distance limit or search radius for an independent-node algorithm is usually more important than with a spreading method, but in any event keeps far distant points from exerting undue influence. If too small a radius is selected for an independent-node algorithm, few points will be selected and the calculations will be based very locally, tending to small closures. On the other hand, a large radius allows many distant points to be used, giving undue influence to the regional pattern and using excessive computer time. Because a spreading algorithm selects points only in the first step, the influence of possibly distant points is reduced.

Some programs have many parameters that can be selected to control the generation of a grid. The user may wish to know if a good combination has been chosen. This difficult problem can involve a combination of geological interpretation and statistical analysis. A point of major importance is the appearance of the contour map that results from a grid. Does the map honor both the specific values and the spirit or trend of the data? Does it have reasonable projections away from the data? Are bends in the contours justifiable? Does the map look reasonable to an experienced geologist?

Cross sections or profiles are also valuable tools for analyzing the grid. These can show features that are not apparent on contour maps, particularly if two or more grids are to be compared. Posting the data values also shows how the grid honors the data. Such displays can point out if the grid is too smooth or if it tends to put "bulls-eyes" around the data points.

Statistical methods can also give an indication of which parameters might be appropriate. Davis (1976), Harbaugh et al. (1977), and Mosteller and Tukey (1977) discuss cross-validation procedures. These procedures delete points from the data, create a grid, and use the grid to determine how well it predicts the deleted points. Using these procedures for several sets of parameters can give an indication of which is best. However, cross-validation should be used with caution. It is only one tool, and it should not be used in place of geological considerations or as more than just an indication. The errors at each point should also be evaluated to be sure that the results are not biased geographically, sampling is representative of the entire gridded area, and the results are not unduly influenced by only one or two points. Kane et al. (1982) also present methods to select optimum parameters. Wren (1975) evaluates a grid in terms of filter theory.

55

Gridding Restrictions

It might be necessary to place restrictions on a grid during generation. For instance, grid intersections distant from data or outside a specified fault block may not be of interest. If these intersections are assigned Null values, the program will ignore this area during later processing; the Nulls have the effect of blanking out part of the grid.

Nulls can be introduced into a grid in several ways. A typical restriction is based on distances from the data locations to the node in question. If a distance limit is used, and no data points (or too few points to calculate a value) are within that distance of the node, then the node should be assigned a Null value. Similarly, all such distant intersections will be assigned Nulls, defining a blank area.

Another way to blank areas in the grid is through data distribution. We may not wish any nodes to be calculated if extrapolation will be required. If so, an envelope or boundary might be formed around the data locations (e.g., a convex hull) and values calculated only inside the boundary, with Nulls assigned to intersections outside. These restrictions can be accomplished with search sectors; generate a Null value if more than a specified number of sectors are empty for a given node.

A third method for areal restriction involves the use of polygons. A polygon is a set of connected line segments, with the last segment connecting to the first (Fig. 4.5). The segments in the polygon are defined by a series of (X,Y) locations. This closed figure defines an area, perhaps corresponding to a fault block, mine pit or advance, or lease.

The gridding program may be instructed to assign values to all nodes within the polygon and Nulls outside it. Alternatively, Nulls may be assigned within the polygon and values outside it. Polygons can restrict the data points that are to be used — that is, only those within a given polygon are used to generate the grid. Polygons are also useful to limit other calculations — to plot contours only within a designated area, for instance.

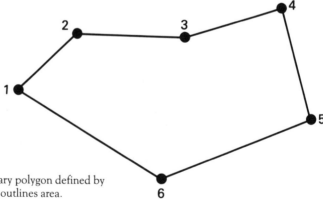

Figure 4.5 Boundary polygon defined by six connected points outlines area.

Another type of restriction limits the minimum or maximum values assigned to the grid. For instance, we might wish to limit the maximum node value to be no greater than a specified value—say, 100. When the grid is made, any calculation that yields a value greater than 100 must be overridden. This adjustment can be made in two ways: (1) the grid is truncated after it is created—that is, the grid is made normally and later all values greater than 100 are changed to 100; or (2) an adjustment is introduced during calculation to cause the grid values to approach 100 gradually. Be aware of how such restrictions might be performed in the program under use.

Triangular Grids

Our discussion has assumed square or rectangular grids. However, alternatives exist. For instance, a grid can be made up of a series of triangles. These triangles are usually defined by the data distribution, a vertex being placed at each data point. The vertices thus correspond to grid intersections, and the value assigned to each vertex is that value observed at the data point. Ripley (1981) discusses theoretical aspects of triangular mesh systems.

The triangular method has the advantages that calculation routines are not needed to assign values to the nodes and that editing can be done easily. In addition, when the map is drawn, the data will be honored—that is, the contours will pass on the correct sides of the points. The triangular method also has potential disadvantages (cf. Harbaugh et al., 1977); many ways exist to connect a series of points into triangles, potentially leading to different maps if the data are sorted differently, and computation to do the triangularization has been expensive.

Watson (1981, 1982), Philip and Watson (1982), and Bowyer (1981) describe methods of forming the triangles in such a way that the triangles are as equiangular as possible. This procedure leads to triangular sets with good properties (Delauney triangles). Bowyer (1981) and Watson (1982) present computational procedures that are said to be efficient, and Watson and Philip (1984) discuss interpolation.

Many advocates of triangular gridding systems stess their advantages over square or rectangular grids, typically presented in terms of mathematical properties (cf. Philip and Watson, 1982; Watson and Philip, 1984). While not arguing with such analysis, we feel that single-surface display is only part of the mapping process and not the entire issue.

In succeeding chapters, we present methods for manipulating grids and data to obtain geologically reasonable maps. Most mapping programs that use square grids can operate one grid against another (chap. 6), allow projections, and so on, but many triangular-based programs cannot. General purpose mapping programs must have features that allow incorporation of geologic interpretation. When surface manipulation capabilities are more widely incorporated into triangular-grid programs, we feel that they will become more widely used.

In order to operate one mesh against another, it is necessary that each has the

57

same triangular framework—that is, the same set of (X,Y) vertices. If a deep horizon has fewer data points than a shallow one, the mesh will not match. In such a case it is necessary that missing points be added to the deeper data. The mesh of triangles is then defined with all these points, and values are interpolated to the added points from the surrounding data. Similarly, each horizon pick will be in a different location for deviated or nonvertical wells, requiring that many extra points be added to the triangular system.

Equivalents to Grids

An alternative to grids is the grid-equivalent, in which the approximated surface is not handled as an independent entity. We have been discussing rectangular or triangular grids that are made up of a set of values treated as a group. The grid equivalent consists of entries in a database. In this case the entries include (X,Y) coordinates of node locations and one or more Z-fields. Each field represents what would have been a grid. The information is organized by location, with many associated variables at each; in contrast, a grid file is organized by variable, with all the nodes representing locations.

As with the grid approach, it is necessary to have a value at each of the calculation locations. Methods discussed above are used for making these calculations. When using the grid-equivalent method, the grid identifier must be replaced by a field identifier, and grid-to-grid manipulations are replaced by calculations between fields, similarly to the discussion in chapter 3. The locations might be in a grid layout, but such a layout is not necessary. However, the program must be able to use the database for plotting contours, and a gridlike pattern is used in virtually every contouring program; interpolations to a finer pattern can be made rapidly (e.g., Davis and David, 1980).

CONTOURING TECHNIQUES

Typical Algorithms

Fewer methods exist for contouring a grid than for creating one. Because of the simplicity introduced by the grid, extremely complicated procedures are not necessary. Two basic steps are required (cf. Sampson, 1975): (1) detection of a contour line intersecting a given grid-square and (2) drawing the line through that grid-square. In the following paragraphs, assume that we are talking about a single contour of, say, value 100; the same process is repeated for every contour.

The first step in contouring is detection of the contour line crossing into or intersecting the edge of a grid-square. This can be done simply by comparing the two grid-node values along a side with the value of the current contour. If the contour value lies between the two node values, the contour must enter the square. The location of the crossing can be calculated by linear interpolation.

If other nearby grid nodes are used, the detection and location of contour entry points at the grid square can be improved. A simple, curved function can be calculated with nodes surrounding the square, and interpolation with this function allows changes in slope to be considered in locating entry/exit points. The exit point of the line through another side is found by the same procedure.

After the entry and exit points of the line have been found, the next step is to draw the line through the square. If a smooth contour is not required, it is possible simply to plot a straight line between the entry and exit points. If esthetic curvature is required, a common method divides the grid-square into several smaller squares or triangles and interpolates intermediate values. The same process is used, but now the straight-line contour segments are shorter and together simulate a curve.

Intersections away from the grid-square under consideration can also be used. For instance, we know the location of the entry and exit points in the grid-square in Figure 4.6. We can also find exit points of the grid-squares on each end of the segment, thereby providing four locations. If a simple function (polynomial or spline) is fit to these points (being sure the line produced passes through the two central points), the central part of the curve can be plotted. A potential pitfall of curve fitting is that the resulting contours may cross.

McLain (1974, 1976), Sutcliffe (1976), and Sibson and Thomson (1981) discuss drawing contours through use of simple, curved functions over the grid-square. Regardless of how lines are drawn through the square, a procedure is applied in turn to each contour line that passes through the square, or the given line is traced throughout the map before going to another contour.

Triangular systems are contoured similarly to a rectangular grid. The sides of the triangles are used to find locations of contours; a contour that falls between two vertex values crosses the side, as with a grid-square. With two sides intersected, it is a straightforward procedure to locate the point of intersection. The contour lines are drawn across the triangle, as described above. Calculation of smaller internal triangles, with interpolated values at the new vertices, increases detail.

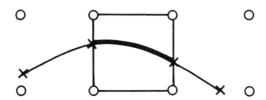

Figure 4.6 Using entry-exit points outside of grid square to fit curve for plotting contour inside grid-square. Dots represent grid nodes, X represents entry-exit points, and heavy line is the plotted contour.

59

Important Parameters and Restrictions

As with gridding, certain program parameters are important in plotting the contour map from a grid. The most obvious one is the choice of contour interval. The interval should allow enough contour lines on the map to define the geographic variation of the surface or variable, but an interval that is too fine clutters the map. As a rule of thumb, for a map that has only large features, an interval of about one tenth to one twentieth of the range of grid values often works well.

Sometimes a constant contour interval is not appropriate. For example, logarithmic variables, such as permeability or sediment grain size measured in mm, commonly show a wide range in values from smallest to largest. With these variables a fixed number of significant digits, rather than a fixed number of decimal places, may be important. With a logarithmic variable, a constant contour interval that covers the entire range of values will show no features in the low-value portion of the map.

Three methods are used to map logarithmic variables. The first converts the variable to logarithms at the data-preparation stage, as is routinely done with sediment analysis on the phi-scale. Gridding these logarithms and contouring the grid gives a map of logarithms of the variable. In this case equally spaced contours are in terms of logarithms. In the second, some programs plot log-contours. The third method uses a grid of the original values. Here the program needs the capability to plot only designated contours; for example, the user might specify that a geometric progression of contour values be plotted (e.g., 1, 2, 4, 8, . . .).

Another consideration for contours involves specifying minimum or maximum values to be plotted. For example, an isochore grid may have projected to negative values. In order to prevent the plotting of negative contours, a minimum contour value of zero should be specified. As another example, the zero contour in a difference grid is plotted to define the subcrop at the edge of a unit (chap. 7).

Another plotting restriction uses polygons. A grid might contain values from edge to edge, but we might wish to plot only a portion of it, say, within a lease or fault block. If the outline of this block is defined by a polygon, it can be used to restrict the plot — that is, contour only those grid intersections within the polygon — and leave blank the region outside the polygon.

Of course, if Null values are in a grid, the contours cannot be drawn in that region. Most programs leave a grid-square blank if any of its four intersections are assigned a Null value.

Printer Displays

Simple contour maps are drawn with the line printer much as they are on the plotter. The page is scaled in the X and Y directions, each print position on the page is found, and the contours are superimposed on these print positions. Rather than drawing the contour as a line, the print positions cut by the line are identified.

These positions are printed with a symbol (e.g., •), and the remainder are left blank. This gives an approximate rendition of the contour line.

Use of the printer saves time in obtaining plots, although the plots are not necessarily easy to use. Distinguishing separate contour lines requires a larger map than is normally needed with a precision line plotter. In addition, if the map is wide, the limited width of the line printer will require that the map be printed in strips and pieced together.

A more convenient way of printer-mapping the information in a grid is the "symbol" or "snapshot" map; Figure 4.7A shows a printer symbol map made from the grid used in Figure 4.3. This map does not have an assignable scale, but gives a picture of the value at each grid intersection. The basic principle here is that each node will be made to correspond to a print position on the page; that is, if a grid contains 30 rows and 50 columns, then 50 print columns in each of 30 lines on the page are used.

Once the grid nodes are made to correspond to the print positions, we relate the nodal values to alphabetic printer symbols. As with plotter maps, a contour interval—say, 125 feet—is chosen. In this case, we find all nodes that have values in

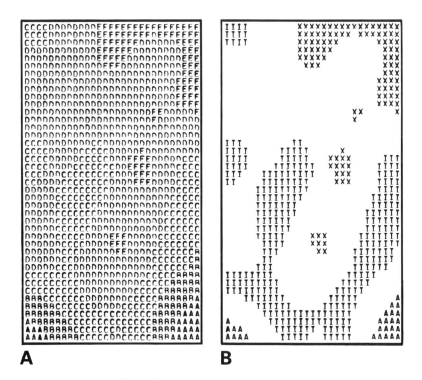

A   **B**

Figure 4.7 "Symbol" map from a line printer. *(A)* Value ranges assigned to printer characters "A," "B," "C," and so on. *(B)* Printer characters assigned to accentuate contrast between value ranges.

61

the range −7625 to −7500. Each of the corresponding print positions might then be assigned "A." Using this procedure, all print positions corresponding to node values from −7750 to −7625 might be assigned "B," the range (−7875, −7750) to "C," and so on.

This method gives a display in which areas having the same printer symbol correspond to grid values in the same range. Whenever two different symbols are in contact, a contour line is implied. For example, the boundary between B and C corresponds to the −7750-foot contour. While not exact, these "lines" give an indication of the pattern of the variable or surface being mapped. Advantages of symbol maps are their speed and low cost.

The choice of printer symbols used can improve the pictorial aspects of the display. Displays are improved if symbols in adjacent ranges vary in density. In the example, assigning the blank character as the symbol for every second range and replacing "C" with "T" and "E" with "X" will show the separate regions more distinctly (Fig. 4.7B). A special symbol should be used for Null grid values.

In chapters 6 and following, techniques are presented that require several grids to be operated one against another. These techniques can be long and complex, and it is often handy to use the symbol maps to check for user errors. Detailed contour maps may not be needed in such applications, and it might be adequate to know regions as being positive, negative, equal to zero, or Null. In this case, three ranges (plus Nulls) would be defined.

Examples and generation of maps by use of the line printer are discussed by Davis (1973), Coppock (1975), Schmidt and Zafft (1975), and Sampson (1975). In addition to showing symbol maps, they include extensions to other printer-based procedures and point out both good and bad features of printer displays.

OTHER DISPLAYS

In addition to drawing contour maps, a grid can be used to create other useful displays. For example, a cross section or profile plots the grid values along a line or connected group of lines, showing detailed variation. The horizontal axis of the plot represents distance along the line of section, and the vertical axis represents depth or elevation. Elevations from a structure grid are derived from the values at nodes along the line of the section. Nonstructural variables can also be used. For instance, a profile can be used to show detailed thickness changes, and a grid of porosity shows porous and nonporous regions along the traverse.

A cross section or profile is valuable for determining if a given grid reasonably represents the data. Several sets of gridding parameters might possibly be appropriate, and displays of alternative choices aid in understanding how data variation is handled by the gridding algorithms. Figure 4.8 shows a cross section with structural grids that are based on the same data but from two different gridding algorithms. The solid line indicates a grid that honors the spirit or trend of the data better than the grid represented by the dotted line.

62

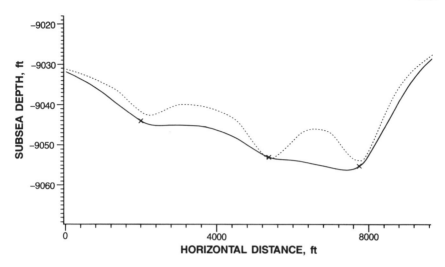

Figure 4.8 Cross section showing two grids based on the same data but different algorithms. The solid line represents a method that better represents the spirit or trend of the data than does the grid from the dotted line. Observed data points are represented by X.

Although cross sections or profiles through a single grid are useful, the primary value of sections is in the display of several horizons or variables at once — for example, stratigraphic relationships between various intersecting horizons. Detailed discussion of cross sections, in particular those involving several horizons in a stratigraphic framework, is found in chapter 7.

Another type of display is a perspective or isometric view of the gridded surface, which provides an impression of three dimensions and is generally more pictorial than a contour map. This display shows the surface or variable as a block diagram; Figure 4.9 is a perspective view of the grid mapped in Figure 4.3. Such displays are not quantitative, but the three-dimensional simulation that results gives a representation of the grid. These displays are particularly useful for people who are not familiar with contour maps to "see" unusual or special features.

The lines in the figure follow the rows of the grid, and the nodal values are plotted as height above some reference. The height variations cause the line spacing to vary, and highs in front hide features in back, thereby creating the visual impression of three-dimensional shape. Techniques for drawing these displays include the use of rows of grid nodes (Fig. 4.9), mesh networks (Fig. 4.10), and obliquely viewed contours, among others. Examples of these displays are given by Davis (1973), Sampson (1975), Schmidt and Zafft (1975), and Ripley (1981).

The user must specify an "eye position" for perspective displays, normally in terms of angles and distances from some point on the grid. Height above the surface affects the distinctness of the features. Viewing direction is also important, as better displays are generally obtained if the surface is viewed updip rather than

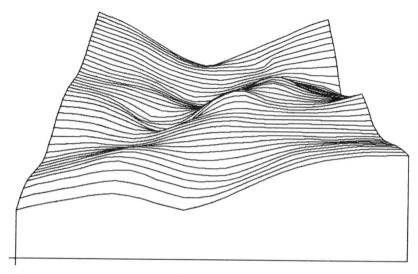

Figure 4.9   Perspective view of surface shown in Figure 4.3, looking toward the south.

down or if extreme highs are placed to the back or side. A program with sufficient flexibility to generate two plots, one with an eye position 2 to 3 inches to the right or left of the other, can be used to create stereo pairs. Trump and Patnode (1960) and Patnode and Hodgson (1964) point out that contour maps can be similarly drawn to generate stereo pairs.

The grid used for contour mapping can also be used for these plots. However, the three-dimensional effect is generally enhanced if the grid interval is made smaller. If the program allows it, a good procedure merely creates a finer second grid from the original by interpolating node values between the original nodes.

Figures 4.9 and 4.10 are simple perspective views. Some programs allow much more information to be displayed on such plots: obliquely viewed contours, well locations, lease boundaries, or other cultural features. Color is used as well. These enhancements require sophisticated programming found only in advanced mapping packages.

Other displays, including variations on contouring, are available to improve the ability to interpret a map. Several methods are based on the concept of a light source shining on the surface, casting shadows in some areas and brightly illuminating others (cf. Peucker et al., 1975; Sprunt, 1975). Tanaka (1950) introduced relief contours by simulating shadow effects through thickening of contours away from the light source. Yoeli (1967) presented "analytical hill shading," which assigns a gray scale over the area, with brightness responding to the orientation of the surface with respect to the light source. This approach, which can provide half-tone displays, is also suitable for perspective views.

If the program at hand does not have capability to plot contours that reflect a hypothetical light source, it is simple to simulate. Create a grid of a plane (see chap.

64

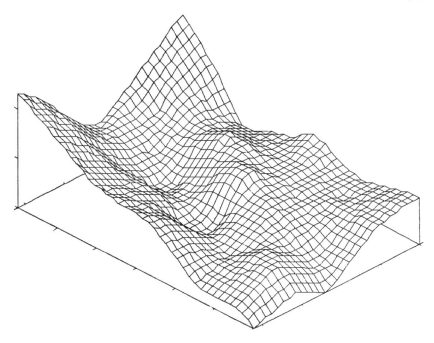

Figure 4.10 Perspective view of surface shown in Figure 4.3, as illustrated with grid mesh, looking toward the southwest.

11) that tilts away from the hypothetical light source. The angle of dip on the plane should be the angle of the light source from the vertical. Add the plane-grid to the grid being mapped (see chap. 6), and plot the resultant grid. The posted contour values are not meaningful, but the contours will show forms. The display is improved if contours are not labeled and the contour interval is made smaller than usual.

CHAPTER 5

Data Characteristics Requiring Special Handling

Some data have attributes that can lead to difficulties in computer analysis and mapping. These data-related problems are not errors in the sense of mistyped numbers, but are conditions of the data that prevent creation of an acceptable grid. We assume the data are already in the computer and that errors associated with data entry, as described in chapter 3, have been corrected. However, even with "clean data" special processing is required for:

1. Mixed data that represent different types of surfaces (e.g., picks for stratigraphic and unconformity surfaces may have been treated as being from the same surface);
2. Data that will not produce an acceptable grid because of areal distribution or variation in Z-values (e.g., large areas without control may exist on the map);
3. Data that contain inconsistencies (e.g., mis-tied seismic lines) or that are from multiple sources (e.g., well and seismic structure picks, or porosity values from cores and logs);
4. Data that imply limits on the variable rather than the exact value at each location (e.g., a horizon deeper than drilling should not be mapped shallower than the bottom of the hole).

The four sections of this chapter describe processing for these data-related problems. The first two sections deal with problems often caused by improper collection or interpretation and sometimes by data distribution. These sections describe how the data should be arrayed in the computer so that mapping techniques discussed in later chapters will operate properly. How the data were initially collected, interpreted, or stored is not important if the information can be put in an acceptable form for mapping.

Two special computer capabilities are used by several of the techniques to be described: inverse interpolation and conversion from a grid to data. These are briefly summarized here and are further discussed in Appendix A. Grid-to-data conversion creates a new data file in which each record contains the X-, Y-, and Z-values corresponding to a node in the grid. Once converted to data, the grid-node values can be combined with other data files, selectively removed or filtered, and used for calculations, just as can other records and fields in a data file.

Inverse interpolation is the reverse of the grid construction process; grid nodes are used to calculate a value at an arbitrary (X,Y) location. The inverse-interpolation process uses the values at nodes near the (X,Y) location of interest to fit a mathematical function. That function is then used to calculate a Z-value at the (X,Y) location of interest.

Inverse interpolation can be performed at any (X,Y) location in a grid, provided some of the grid nodes surrounding the location have assigned values (i.e., are not Nulls). Extrapolation outside the area of a grid is normally not possible in inverse interpolation. The (X,Y) locations typically used are input from a data file, and the interpolated values are written onto that data file as a new field or onto a new file.

SEPARATING STRATIGRAPHIC PICKS FROM UNCONFORMITY PICKS

An error that is commonly made during data collection is mixing stratigraphic picks with unconformity picks. A stratigraphic pick is the intersection of a borehole with a litho-, bio-, or chronostratigraphic horizon. Unconformity picks represent the intersection of a borehole with an unconformity. *Rock-top* is a term that encompasses both of these and may represent either a stratigraphic horizon or an unconformity. For mapping by hand or with the computer, it is important that stratigraphic picks and unconformity picks be handled separately (Iglehart, 1970; Jones and Johnson, 1983).

Figure 5.1A shows that incorrect grids can result from mixing these two types of data. Here the geologist used a mixture of stratigraphic and erosional tops for horizon B. Surface C is an unconformity that truncates surface B and removes a portion of rock unit 1. In well 3 the elevation at the top of unit 1 was entered for both unconformity C and stratigraphic horizon B. Gridding data from horizon B produces surface D, which deviates drastically from the true geologic surface B.

These two types of surfaces should be identified and mapped separately and then combined to form the complete top of the rock unit (Fig. 5.1B). Surface B has been eroded in well 3, so no pick exists for that surface, and a Null value is recorded. Regridding produces a reasonable projection. Surface B now projects

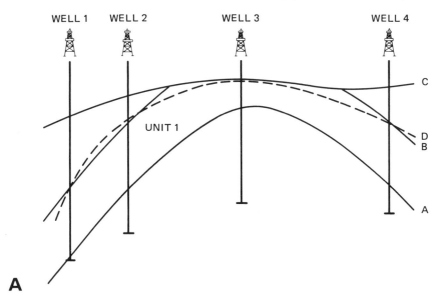

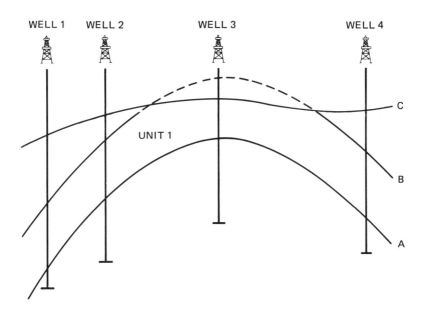

Figure 5.1 Cross section showing conformable surfaces A and B truncated by unconformity C, with portion of rock unit 1 removed. *(A)* Surface B does not exist in well 3, but has been picked as the top of rock unit 1. The resulting grid, surface D, is incorrect. *(B)* The pick for surface B in well 3 is set to Null and correctly produces grid B. (After Jones and Johnson, 1983, p. 1419)

69

above unconformity surface C into areas where it should not exist; this is accept-able for the moment because techniques are available for removing the projection (chap. 6).

Data will sometimes have already been compiled with stratigraphic and uncon-formity picks recorded for the same surface. If the dataset is small, the geologist can edit it, separating the picks that were incorrectly identified as the same. However, it is possible to automate portions of the process by comparing fields for adjacent surfaces to determine if one or the other is in error. This can be done by a program or by a series of operations on data fields (chap. 3) and can be tailored to each geologic situation.

To aid in processing, whether by hand or with a computer program, a list of all stratigraphic horizons and unconformities should be made. Order the list with the shallowest (youngest) surface on the top and the deepest (oldest) on the bottom. This list is used as a reference for manual modifications and is a guide for setting up program steps or data operations.

The goal is to separate stratigraphic and unconformity picks that have been incorrectly identified, so three aspects need to be checked: (1) all picks should occur in the correct stratigraphic order; (2) the same elevation should not be assigned to more than one pick; and (3) unconformity picks should not be identi-fied as data representing stratigraphic surfaces. Each of these aspects is described in detail below.

Correct Stratigraphic Order

All picks are in the correct stratigraphic order if the elevation values in fields representing younger (shallower) surfaces are higher than older surfaces. (Discus-sions here assume that the stratigraphic units have not been overturned.) When a pick for a surface is found to be deeper than an older surface, the record should be identified in some manner for later evaluation and correction. No modifications can be made automatically by the computer program, as it is impossible to predict which pick is wrong or what the correct value should be.

Multiple Picks at the Same Elevation

Another error that is sometimes made when collecting data is assigning the same elevation to more than one pick. Two situations commonly occur: (1) record-ing the same pick as both an unconformity and a stratigraphic pick if the top of a rock unit is eroded and (2) recording the top and base of a rock unit as the same elevation when the rock unit has pinched out.

The first situation is a version of mixing stratigraphic picks and unconformity picks. The geologist has recognized that the pick is at the unconformity, but has also recorded the pick as the stratigraphic horizon bounding the top of the rock unit even though it represents only the rock-top of that unit. In this case the value entered as the unconformity pick should be used and the other set to Null. If the same elevation value has been entered for more than one pick, the stratigraphically lower pick(s) is usually set to Null.

In the second situation, it may seem reasonable for a geologist to record coincident picks for stratigraphic horizons bounding a unit where that unit has pinched out (Fig. 5.2A); however, problems may arise during gridding and contouring. In the area where the unit is missing, grids built from these coincident picks will not have identical values (Fig. 5.2B). It is pointed out in chapter 9 that a map of the unit will have zero thickness over much of the missing area but can contain small positive and negative thicknesses. This situation arises because the bounding grids are built separately and have no "knowledge" of their geologic relationship.

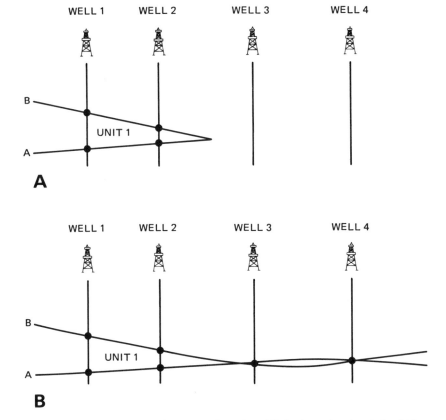

Figure 5.2 Cross section showing surfaces A and B that bound rock unit 1, which pinches out between wells 2 and 3. (A) Picks for surfaces A and B have been made coincident in the area of pinch-out so that rock unit 1 is forced to have zero thickness in those wells. (B) Grids A and B built independently; because of differences in extrapolation, they will not be coincident throughout the area of pinch-out.

71

A solution to this problem is to record tops for only one of the horizons and allow the other to project past the pinch-out, as in Figure 5.1*B*. This projection can then be removed using techniques described in chapter 6. If the problem is not solved, a value slightly higher than the original can be entered for the lower surface, with the upper surface retaining the correct value. This will force the grids to cross in the area where the unit pinches out. Similarly, the upper surface could be projected below the lower surface. Chapter 9 discusses a variation of this technique and several others dealing with pinch-outs, truncations, and baselaps and the location of their associated zero-thickness lines.

Separation of Unconformity and Stratigraphic Picks

Often data will already have been collected and unconformity and stratigraphic picks identified for the same surfaces. The geologists can rework a few data points, separating picks that were incorrectly identified as the same. For a large amount of data, however, manual editing may be a monumental task. Computer programs can be written to speed portions of this task. Unfortunately, the entire process cannot be automated because some geologic interpretation is required. The following discussion identifies common problems with unconformity and stratigraphic picks and sets out procedures for correcting those problems. The procedures can be done manually or with the computer.

Stratigraphic and unconformity picks have probably not been mixed if (1) every unconformity has an elevation value, (2) all the picks are in correct elevation order, and (3) no picks have the same elevation. In this case, Null values correspond to stratigraphic horizons that have either been eroded or were not deposited. However, there is a possibility that stratigraphic and unconformity picks have been mixed if one or more unconformity fields have Null values.

If a Null occurs at an unconformity, yet stratigraphic horizons occur between that unconformity and the next-younger (shallower) unconformity having an elevation value, then something is wrong. The usual reason for a missing unconformity pick is removal by later erosion. In that case, none of the surfaces between those unconformities should be present (i.e., their fields should all be Nulls). If surfaces between these unconformities have values, then the lower unconformity is probably present but has been incorrectly picked as a stratigraphic top. For example, Figure 5.3 shows a cross section with conformable surfaces A, B, and C truncated by unconformity D and with surface E lapping onto D. Well 1 penetrates stratigraphic surfaces E and A, and unconformity D, but does not cut surface B.

Table 5.1 indicates the information recorded for the wells in the section. Some of the data are incorrect. Although well 1 penetrates unconformity D and not surface B, an elevation value was entered incorrectly for surface B and Null for D. The surface most likely to have been misidentified is the first stratigraphic horizon below the missing unconformity. If this is the case, that horizon should be set to Null and the unconformity assigned that elevation value.

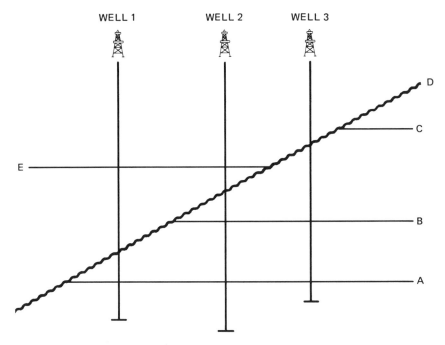

Figure 5.3 Cross section showing conformable surfaces A, B, and C truncated by unconformity D, with surface E lapping onto D. Picks in wells penetrating these surfaces should be recorded only for those surfaces penetrated; all others should be assigned Nulls.

It is possible that the stratigraphic horizon beneath the missing unconformity was correctly picked as in well 2 (Table 5.1) but the unconformity pick was incorrectly omitted or could not be identified. In that case, the unconformity lies somewhere between the highest surface in the stratigraphic sequence below the missing unconformity (surface B) and the lowermost surface in the sequence above the missing unconformity (surface E). Elevations for these two surfaces should be used as constraining limits when the unconformity surface is constructed. Techniques for this procedure are presented in the last section of this chapter.

The approach just described will isolate most problems; however, if an unconformity field and all surfaces in the well above that unconformity contain Nulls, it is possible that the last horizon with an elevation value is really an unconformity (Table 5.1, well 3). In most cases the pick has probably been correctly identified, but there is no automatic method for determining whether it has.

In a regional study some unconformities may grade laterally into conformable surfaces. These surfaces are handled in a variety of ways. The easiest treats them as unconformities everywhere. When this is done and the above procedures are applied, Null values for these surfaces may be identified as potential problems

73

TABLE 5.1
Information Recorded for the Wells and Surfaces in Figure 5.3

Surface	Well 1	Well 2	Well 3
E	value	value	Null
D[a]	Null[b]	Null[b]	Null[b]
C	Null	Null	value[b]
B	value[b]	value	value
A	value	value	value

[a]unconformity
[b]incorrect or potentially incorrect data

when they are actually correct. The advantage is that errors will not be overlooked. Another method for handling these unconformities is to divide the data into subsets, one set representing the area where the surface is an unconformity and the other the area where the surface is conformable. Each dataset is then processed using a slightly different geologic interpretation. This extra effort is usually not warranted.

SUPPLEMENTAL DATA FOR GRID CONTROL

Sometimes data will be too sparse or irregularly distributed to produce an acceptable grid. Such situations may arise for a number of reasons: (1) little data are available in the area of interest; (2) the geologic feature of interest is small, so only a few holes have penetrated it; and (3) the section is thick, with deeper horizons penetrated by only a few holes.

Sparse or poorly distributed data may produce a grid that is unacceptable, primarily because of uncontrolled extrapolations. In these situations, additional control may be added by incorporating more data in the form of control points, either automatically or manually. Automatic techniques are appropriate for projects where new data will continually be added and human intervention would slow the process. Each time new data are added, the entire dataset is reprocessed and a new set of automatically generated control points is created.

Manual addition of control points is tedious and costly, but it is often required in empty areas to control the shape of geologic features that are small relative to the area being mapped. These small features do not contain enough data to build a grid directly or to guide an automatic-control-point technique. Stream channels, reefs, units that pinch out, and steeply dipping fault planes are features that are commonly restricted to a small portion of the map area, and they sometimes require additional control points.

Automatic Addition of Control Points

Although probably as many automatic procedures as mapping programs exist, most have some basic similarities. They typically add control points to areas having no data and use the general trend of the existing data to estimate values for those points. Two common procedures use ring points and coarse grids.

Ring-Points Procedure

The ring-points procedure calculates values at points around the edge of the map. These points are then added to the original data and used to generate the final grid. The ring-point procedure is used to control surfaces that must be gridded beyond the edge of the data to the edge of the map. For example, deep surfaces may be penetrated in only a few areas, and therefore data coverage is significantly less than that for shallower surfaces. If structure maps covering the entire area are desired for every surface, grids for those lower surfaces must extend as far laterally as do the shallower surfaces. Grids of the lower surfaces will be poorly controlled near their edges.

Few ring points are required to control the gridding process; eight points spaced evenly around the grid perimeter are usually sufficient. Several techniques can be used to assign values to the ring points, but common ones include trend surface, local average, and average value.

In general, a trend surface is a simple function of geographic coordinates that passes as closely as possible to all data points while still honoring mathematical constraints (chap. 11). The typical constraint placed on a polynomial trend-surface equation is the order of the trend. A first-order trend surface is a plane tilted in space; it usually approximates the surface well enough for this application. Values are interpolated from the function at the locations of the ring points.

Local-average methods use data points in the area around a ring point, calculate an average for those points, and assign that value to the ring point. Since most programs do not have a procedure to do this automatically, an approximate method uses the gridding algorithm and a very coarse grid interval. For many algorithms, several points falling within one grid-square are essentially averaged when assigning a value to the grid nodes. The coarse grid therefore approximates a local average of the data values.

The average-value method averages all original data points and assigns that average value to the ring points. This method is appropriate for flat-lying surfaces, isochores of relatively constant thickness, or variables without trends, while trend and local average techniques are appropriate for dipping or irregular surfaces. If the gridding algorithm used to build the final grid uses data outside the grid limits, it may be appropriate to move the ring points some distance beyond the grid perimeter. This tends to produce more appealing contours at the edge of the map.

Coarse-Grid Procedure

The coarse-grid procedure is most appropriate for "spreading" gridding algorithms—that is, those that calculate values at grid nodes around data points and then use those values to calculate the nodes farther from the data. These algorithms tend to extrapolate uncontrollably away from the data. The coarse-grid procedure helps control this extrapolation. This procedure is sometimes also appropriate for other situations in which a strong trend must be forced into the final grid.

The coarse-grid procedure uses the following steps: (1) Build a grid with an interval several times (usually four or five) larger than the interval required for the final grid; (2) convert the coarse grid to data (grid-to-data); (3) edit the data file, deleting those records that are near original data values; (4) combine the coarse-grid data with the original data; and (5) regrid the "enhanced" data using the desired final interval.

Construction of the coarse grid may use either a standard gridding algorithm or a trend surface. The coarse-grid interval tends to average data points that occur within the same grid square, producing a moving-average effect that smooths smaller features while keeping larger ones. Sharp local gradients are therefore subdued, as are extrapolations caused by these gradients. Also, because of the coarse interval, large areas without data will contain few rather than hundreds of grid nodes. With fewer grid intersections and more subdued gradients, the extrapolations are less likely to get out of hand.

Once the coarse grid is built, it is converted back to data. This results in a new dataset with three fields—X, Y, and Z—and each record corresponding to a node in the coarse grid. The new coarse-grid data is compared to the original data, and any coarse-grid records within a specified distance (usually two coarse-grid intervals) of an original data point is either removed or set to Null. The modified coarse-grid data is combined with the original data, and the final grid is built with a grid interval appropriate for the original data distribution.

The coarse-grid procedure is useful for irregularly distributed data containing large open areas but is of no value for evenly distributed data since all coarse-grid data close to an original point would be removed. This procedure is completely automatic in some programs and requires the user to specify only the intervals in the coarse and final grids.

The automatic technique to be used depends upon the surface structure and data distribution. Some experimentation is required to recognize when control points are needed and which technique to use. A combination involving both ring points and a coarse grid has been found useful in many situations.

Manual Addition of Control Points

Manually derived control points ("dummy points") should be used only as a last resort and applied conservatively when used. Using such control points forces a shape on the surface that, although perhaps geologically sensible, is not based on data values. Most projects are ongoing and new data are constantly being added, so there is always a chance that previously used control points will conflict with new data. Because control points require constant updating with such projects, the fewer and more systematically distributed those points are, the easier they will be to maintain.

Each situation that requires manually added control points is unique, and geologic interpretation should control the locations and values of the points. Nonetheless, the authors have found some techniques or guidelines useful in working with manual control points.

1. Use the minimum number of points possible to define the geologic feature. Fewer points are less likely to conflict with new data, and if they do conflict, they will be easier to edit.
2. If possible, distribute the control points systematically. If problems arise in a particular area, the points near that area are more easily found.
3. Use geological judgment to establish the location and value assigned to each point.

Again, we recommend against the use of manual control points. If they must be used, it is best that a sound geologic basis be established for their use. For instance, consider the movement of a tracer in groundwater from a point source. If samples were not taken in front of the leading edge of the tracer plume, the observed values would be projected without control outside the affected area. Control values with zero (or the background value) could be inserted, but where should they be located? Knowledge of when the tracer was injected and of maximum flow rate could be used to locate control points beyond the outermost position of the plume.

MERGING DATA FROM DIFFERENT SOURCES

Different Data Sources

Contour mapping often requires that two or more sets of geologic data, each derived from a different source yet purporting to represent the same surface or variable, be combined. If one of the datasets contains errors or deviates systematically but is combined with the other data and contoured, the map from the combined set may be less realistic than maps contoured independently from each of the other sets.

Systematic error can be found in geophysical data where maps of seismic, magnetic, or gravity surveys show mis-ties at line intersecions. Mis-ties in seismic data may be caused by different processing techniques, lines of different vintage or origin, mislocated lines, dipping horizons, inconsistent interpretation of reflec-

77

tions, small discrepancies in digitization, or combinations of these. Potential field measurements can be affected by diurnal variation or instrument drift.

Figure 5.4 shows a contour map generated from two sets of mis-tied seismic lines. Gridding and contouring this data exaggerates line mis-ties into "clover-leaf" anomalies that mask the true structure. When the east-west and north-south sets of lines are contoured separately (Fig. 5.5A, B), the true structure is exhibited. However, each set contains details that the other does not, and using only one throws away useful information. We therefore want to extract maximum information by combining the line sets; thus, we need a means for correcting the mis-ties.

Data-merge is an important tool for merging two datasets that contain errors or shifts. The error between the two datasets is calculated and all of this error is removed from one of the datasets or a portion of the error is removed from each. This is done by shifting the Z-values of one or both datasets so the data can be gridded and contoured correctly after being combined.

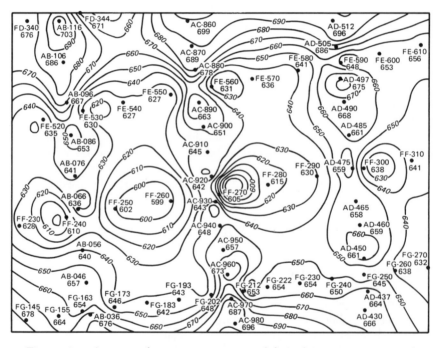

Figure 5.4 A structural contour map generated from data containing mis-tied seismic lines. The mis-ties produce anomalous structures.

Figure 5.5 Separating the east-west lines (A) from the north-south lines (B) and contouring separately depicts the true structure correctly. However, each map contains details the other does not.

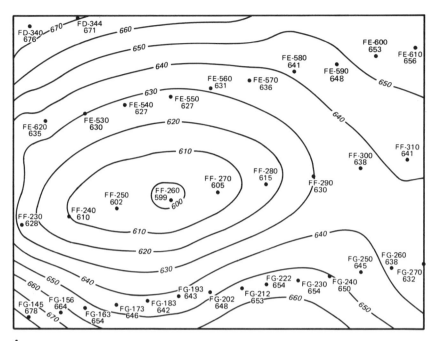

A

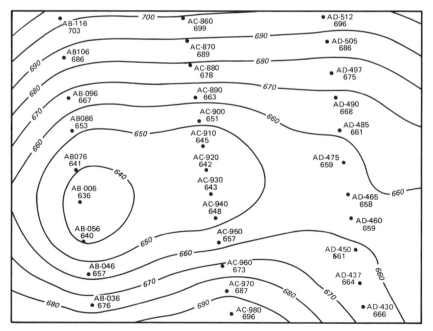

B

79

Data-merge was applied to the mis-tied seismic lines described above. The north-south lines were grouped together as one dataset and the east-west lines as another. All of the error was assumed to lie in the north-south lines, so their Z-values were adjusted to match the east-west lines. Figure 5.6 shows a contour map of the combined datasets after modification. In addition to the more realistic contours caused by removal of the mis-ties, the final map contains more detail than either of the separately contoured groups of seismic lines.

Foster et al. (1970) describe a method similar to Data-merge that corrects for line-tie errors at intersections of aeromagnetic survey lines. Their method also separates the lines into noncrossing sets. Errors are calculated at all intersections of the two line sets, and a statistical model is used to represent smoothly varying error over the entire length of each line. Each line set is then adjusted to remove half the error, thereby tying the lines.

The Data-merge procedure can be applied to a variety of other geologic data problems. For example, it is common in petroleum exploration for a geologist to contour structural elevations derived from both well logs and interpreted seismic profiles. Often an area has extensive seismic coverage but comparatively few wells. The seismic data, converted from travel-time to elevation with seismically derived velocities, are closely spaced along the seismic lines so that details of structure and

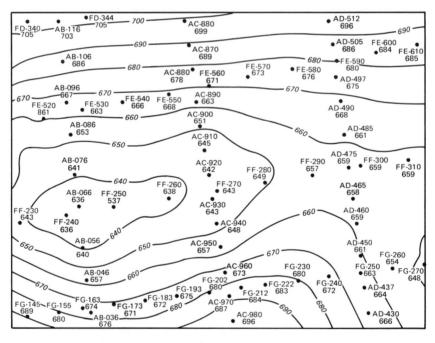

Figure 5.6 The structural contour map after Data-merge was applied shows details from each line set and contains no mis-ties.

stratigraphy are present (Fig. 5.7A). On the other hand, the widely separated wells show only broad features when contoured (Fig. 5.7B).

Besides representing different degrees of detail as a result of spacing, these two types of data differ in accuracy and precision. The well data accurately record the elevations, while the interpreted seismic picks may differ (perhaps systematically) from the true horizon by several hundred feet. Combining the datasets into a common "pool" without adequate adjustment can produce maps containing spurious anomalies that may obscure significant features. The map in Figure 5.7C

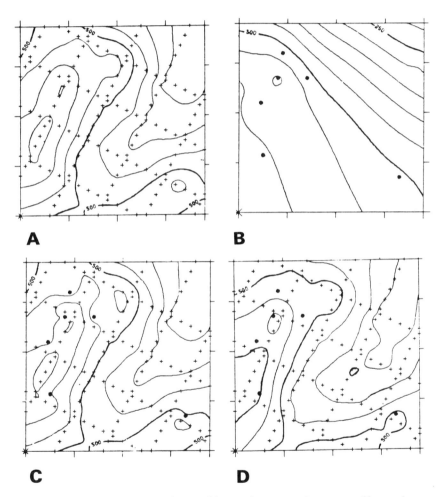

Figure 5.7 Combined data from well logs and interpreted seismic profiles produce a structural contour map. (*A*) Seismic data contoured separately. (*B*) Well data contoured separately. (*C*) Well and seismic data combined and contoured without adjustment. (*D*) Well and seismic data combined and contoured after applying Data-merge.

81

needs extensive modification to be acceptable, especially in the vicinity of the northernmost wells. However, Data-merge removes the error from the seismic data, shifting Z-values to match the well data and thus making the two data types more compatible (Fig. 5.7D). The resulting map honors the accurate well-log picks and contains the superior structural detail of the seismic data.

Another problem common to petroleum production occurs when reservoir porosity values are derived from two sources: core analysis and petrophysical logs such as induction, sonic or neutron logs. Nearly all current wells are logged, so porosity values can be calculated for any reservoir penetrated. Porosity data from core analysis typically are available for only a few wells, although these are likely to be accepted as the more accurate porosity measurements. The distribution of log-calculated porosity probably will not match that of core porosity.

Suppose we use core-porosity values as being more reliable at those locations where both types of porosity measurements are available. The porosity map generated from the log-derived values and the substituted core-derived data may contain anomalies in the vicinity of the substituted values. To produce a more satisfactory map, the geologist would have to correct log-derived values subjectively to make them agree with nearby core-derived values. Data-merge is used to remove the error between the two porosity types—that is, shift the log-derived values to match those from cores.

The amount of manipulation required to correct one or both datasets by hand in these examples is laborious and time-consuming. Data-merge is an effective tool for automating these manipulations without sacrificing data integrity.

The Data-Merge Procedure

The Data-merge procedure is outlined diagramatically in Figure 5.8, following the seismic example described above in which the east-west lines were honored and the north-south lines were modified to remove the mis-tie. The figure shows a cross section passing through data points along one of the north-south lines. Several east-west lines cross this line and, as they are normal to the plane of the section, appear as single points. The north-south lines must be shifted to match the crossing east-west lines, but without losing the detailed shape information they contain. The process includes five steps:

1. Grid one of the datasets (Fig. 5.8A);
2. Calculate the difference (residual or error) between the nongridded dataset and the grid at each data location (Fig. 5.8B);
3. Build a smooth grid of these differences (Fig. 5.8C);
4. For each point of both datasets, determine the difference from this smooth residual grid and shift values of one or both datasets by all or a portion of that difference (Fig. 5.8D, E);
5. Combine the shifted datasets into a single merged data file.

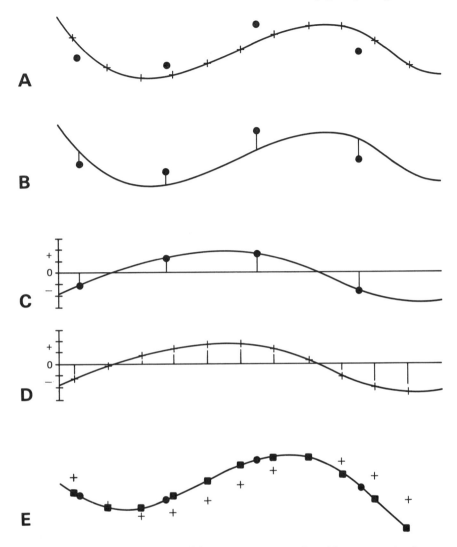

Figure 5.8 The major steps of the Data-merge procedure. The cross section lies along a seismic line (pluses represent data along the line). Several cross-lines (data represented by dots) cut the section at approximately right angles. (*A*) One of the datasets (along line) is gridded. (*B*) Differences between the other dataset (cross-lines) and the grid are determined. (*C*) A smooth grid is built of the differences. (*D*) Values at all points for both datasets are calculated from the difference grid. (*E*) One or both sets of original data points are shifted by the difference; squares represent shifted data values.

83

The following discussion adds details to these general steps of the Data-merge procedure.

Step One: Grid the More "Areally Comprehensive" Dataset

Geological datasets are rarely equal in total number of data points, size of the area covered, and uniformity of areal distribution. That dataset providing the most extensive and uniform coverage is referred to here as dataset A; the other set is referred to as dataset B. Specifying a dataset as being either A or B has no relation to the "error" in that dataset, but only to which will make the more reliable grid.

When building the grid of dataset A, use a grid interval and other parameters that will create the best grid for that dataset. Do not be concerned about whether these parameters are acceptable for dataset B. However, the grid should extend far enough laterally to encompass all points in dataset B and should be reasonably well controlled, even at the edges.

In our example, the east-west seismic lines were defined as dataset A and when gridded produced the contour map in Figure 5.5A.

Step Two: Calculate Differences Between the Two Datasets

Use inverse interpolation with the grid of dataset A to calculate values at the locations of B. Incorporate these interpolated values into dataset B as a new field. Subtract the interpolated B values from the original B values, placing the difference or residual in another field.

In our example the north-south seismic lines were interpolated on the grid from dataset A, and residuals were calculated by subtracting the interpolated values from the real values.

Step Three: Build a Smooth Grid of the Residual Values

The grid of residuals must be generated in such a way that it contains only the trends of the residuals and extends as far laterally as do the A and B datasets. Usually fewer data points are available for building the residual grid than for building the A grid. Since the residual grid is used to shift the datasets, any incorrect extrapolations and interpolations will be incorporated in the final merged dataset. A polynomial trend grid (chap. 11) could be built from the residuals and used for the smooth grid. However, some experimentation is needed to determine the appropriate degree of polynomial.

Standard considerations for choosing a grid interval do not apply when gridding the residual data values. If the residual points are widely spaced, a grid interval appropriate for the residual data will produce acceptable Data-merge results, provided extrapolations are not a problem. However, if the residual data are clustered and a grid interval small enough to honor all points in the cluster is used, the detailed information of the cluster would be removed when the residual is subtracted from the B dataset. A coarse residual grid does not honor any of the clustered B data points but passes through their average position. Subtracting this residual from the B dataset shifts all of the clustered points by about the same amount, and their position relative to each other (detailed shape of surface B) is preserved.

A general approach for building the residual grid is to (1) use a grid interval five to ten times that used for building the dataset A grid and (2) incorporate additional control points around the edge of the map. The coarse grid interval avoids the problem of clustered data because multiple data values within the same grid square are commonly averaged by gridding algorithms. By reducing the number of grid nodes between distant data points, the coarse grid interval also reduces the likelihood of wild interpolations, common in some gridding algorithms. The added ring points reduce the extrapolation problem by keeping the grid under control near the edge. Ring points may not be needed for a low-degree polynomial. However, high-order polynomials will require ring points and may still extrapolate wildly near the edges.

Few ring points are required for grid control; the authors have found eight points, two on each side of the grid, to be sufficient, although this can depend on the gridding algorithm used. These ring points should be placed slightly outside the area of interest. If ring points are placed outside, the residual grid is not particularly sensitive to the Z-values assigned to them. In many cases residuals fluctuate about zero, so zero is a good value to assign to ring points. If the approximate magnitude of the residual is known (e.g., a datum shift or constant error), this value can be used at ring points to obtain greater accuracy.

The use of ring points strengthens the otherwise powerful Data-merge capability. Had dataset B consisted of only a single line, very few points, or any other limited distribution that would otherwise be "ungriddable," a residual grid could not be generated, and the merge process would be impossible. However, supplementing the ungriddable data with ring points allows us to create a residual grid, and Data-merge can be successful.

In our example the residual values from Step 2 were gridded with an interval ten times that of the grid of dataset A, producing a relatively smooth residual grid. To control the extrapolation at its edges, eight ring points were introduced, each with a value of zero.

Step Four: Use the Residual Grid to Correct the Dataset(s)

Application of Data-merge shifts Z-values in one or both datasets so that when combined they will produce a reasonable grid. The decision as to which dataset should be honored and which shifted is quantified as a shift ratio, through assignment of weights that sum to 1. For example, if dataset A is to be honored and B shifted, then the shift ratio A:B will be 0:1. All of the corrections are applied to dataset B, and dataset A is not modified. If each dataset is to be weighted equally, then the shift ratio will be 0.5:0.5; half the residuals are assigned to each dataset. If A is to be weighted 80% and B 20%, then the shift ratio will be 0.2:0.8.

Use inverse interpolation to calculate values from the residual grid at the locations of datasets A and B. Incorporate these interpolated values into their respective datasets as new fields. Multiply the values in the new residual field by the weight from the shift ratio. If the shift ratio is 0.2:0.8 (honor dataset A 80% and B 20%), then the A residuals are multiplied by 0.2 and the B residuals by 0.8. Subtract the modified residual values from the original values, creating a new field in each dataset that now contains the corrected values.

In our example, data values in both the A and B datasets were interpolated on the residual grid, and the interpolated residual values were incorporated as new fields in the respective datasets. Since the north-south lines (dataset B) were to be shifted to match the east-west lines (dataset A), the shift ratio was 0:1. The interpolated residual values in the A and B datasets were multiplied by 0.0 and 1.0, respectively. The resulting values were then subtracted from each dataset's original seismic values. In this example, the values in the A dataset were not modified.

Step Five: Combine the Corrected Datasets into a "Merged" Dataset

The X-, Y-, and corrected Z-fields from both datasets can now be combined. Typically, we will also want to put the following into this merged dataset: identification fields for distinguishing the A and B data records, fields for other surfaces, the original data on which the Data-merge was performed, and the interpolated residuals. By retaining this information, we no longer need the original datasets, although they can be regenerated at any time. If the shift weights need to be changed, all of the information required to do so is in the data file.

In our example, shifted dataset B was combined with unchanged (honored) dataset A. Comparison of the grid made from this merged data (Fig. 5.6) with the original grid (Fig. 5.4) shows the effectiveness of Data-merge. The final map is not only an improvement over the unmerged version but is superior to the maps of the individual line sets (Fig. 5.5A and B) because it incorporates details found in each of them.

Other Data-Merge Considerations

In the merged data file, a point from dataset A may be located near a point from dataset B. Because the Data-merge process involves interpolation and extrapolation, points from one dataset may differ from those in the other dataset by a few vertical units, leading to two possible difficulties: (1) Points from each dataset are in the same grid square and differ by a few vertical units, so their values may be averaged or one honored in preference to the other; and (2) if those points are near but in neighboring grid-squares, the gridding algorithm may create a gradient between the points while trying to honor both, producing high and low values on opposite sides of the points.

Removing one of the points will not drastically change the structure but will eliminate the false extrapolation. This may seem radical, but differences still remaining between the two points are normally small relative to the surface structure. One of the datasets will typically have been honored 100% and the other 0%, in which case the point would be removed from the nonhonored dataset. If the datasets were weighted equally, the decision would be arbitrary or dependent on other factors unique to the particular situation.

Removing too-near points can be a manual editing process but is more conveniently done by the computer. The best approach is to compare all points in the dataset from which points are to be removed to all points in the dataset that is to be left unaltered; any point in the dataset to be modified that is within 1.5 grid intervals (the interval used for final mapping) of a point in the other set would be changed to Null.

In our example, the map constructed from the combined Data-merge data file showed no problems related to points being too close together.

A third set of seismic lines could cross the two previously merged. To merge this set, treat the previously merged set as dataset A and the new set as dataset B. Repeat as many times as necessary to combine all sets into one file.

The Data-merge process, when expanded into detailed steps, seems complicated and perhaps not worth the effort to master all of its details. The authors have found that specifying the program instructions for the first time is indeed an onerous task. However, once the Data-merge procedure has been worked out, adapting that "model" to other applications is easy, and Data-merge becomes a valuable mapping tool. With a little programming effort, the Data-merge procedure can be built into a single mapping task that requires only a few user instructions, thus significantly reducing the expertise required of a user.

DATA THAT REPRESENT LIMITS

Mapping problems arise in which some data points cannot be used directly for creating a grid, but the points nevertheless contain information that must be honored or taken into account. The classical example from the petroleum and minerals industries involves holes that are not drilled to the surface being mapped. Boreholes that are not drilled deeply enough to penetrate the surface could not be used for gridding and mapping. However, relative sparseness of control might allow the grid at some locations to project upward and attain elevations shallower than the known base of some holes.

Figure 5.9 illustrates the problem in cross section. Here boreholes 1, 3, and 5 penetrate to the surface (solid line), but holes 2 and 4 do not. The grid that results from use of only the three points is indicated by the dashed line. The grid is acceptable at hole 2 because it passes below the base of that hole. However, the grid

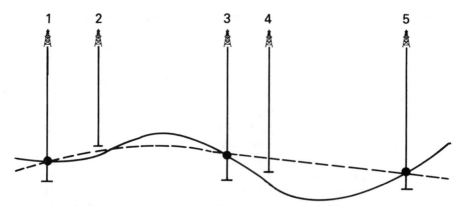

Figure 5.9 Cross section showing structural surface (solid line) and five boreholes. Holes 1, 3, and 5 have elevations picked for the surfaces, but the others do not. The dashed line represents the grid generated with the wells containing picks. Hole 4 is violated by the grid.

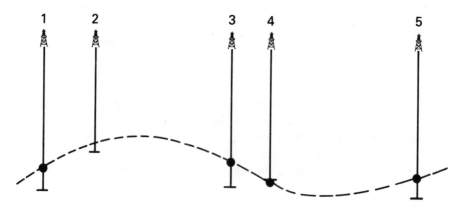

Figure 5.10 Same cross section as in Figure 5.9, but now a fixed pick has been placed at the base of hole 4. The dashed line represents the grid generated with the wells containing picks. Hole 2 is now violated by the grid, and another iteration is required.

is shallower than the known bottom (or total depth, TD) of hole 4. Although this hole cannot be used to create the grid, it can be used to indicate that the grid is in error at that location.

Jones and Jordan (1975) presented a procedure to use in this situation. In Figure 5.9, holes 1, 3, and 5 penetrate the surface, and therefore its elevation is known at these holes; call these known values fixed picks. On the other hand, holes 2 and 4 put constraints on the grid, as the elevations of the bottom of the holes define limits. Our goal is to create a grid that correctly honors the known fixed picks, but at the same time the grid must honor all limits in the other holes.

TABLE 5.2
Example of Input for Data Containing Limits

X	Y	ID	Z	CODE	
—	—	1	−1930		fixed pick
—	—	2	−1920	BELOW	short hole
—	—	3	−1922		fixed pick
—	—	4	−1931	BELOW	short hole
—	—	5	−1927		fixed pick

When the Z-values are submitted to the computer, fixed picks are entered in a field normally. The data that define limits must also be entered. These limits could be placed in another field, but it is more convenient to put the limiting value – that is, the elevation of the bottom of the hole – in the same field. A second data field or modifier is necessary to indicate that this data point is a limit. For convenience, leave this second field blank for data with fixed picks, and use a code such as BELOW to indicate the presence of a limit.

Table 5.2 shows recorded data for the five holes in the figure. Here the (X, Y) coordinates are not listed but would normally be included. Field Z contains elevations of the fixed picks and limit values. Field CODE indicates the type of data point. A brief comment is also included, although it would not be put in the data records.

Briefly, the gridding procedure begins by using only the fixed picks to create a grid. This grid is then compared to the limits. If a constraint is honored (i.e., the grid is deeper than TD), there is no problem at that hole. However, if constraints are violated, the limits are converted to fixed picks and combined with the others, and the grid is regenerated. The new grid is then tested against the remaining constraints.

This procedure therefore involves an iterative process, though one iteration is usually adequate. Figure 5.10 shows the holes in Figure 5.9. However, as a result of the first iteration, a pick is specified at the base of hole 4. When the grid is now generated using these four points (the dashed line), an inconsistency results at hole 2. The grid projects to a higher value at this location than previously because correcting the grid at hole 4 caused a problem at hole 2. It is thus necessary to make another iteration – that is, add the base of hole 2 to the dataset and regrid. Although this complication is not common, it is necessary that the gridding/testing process be repeated until there are no further violations.

The entire process can be done wholly within a mapping program, using data calculations, gridding, and inverse interpolation, or as a series of steps between the mapping program and a data manipulation program that tests for violations. The detailed steps are as follows.

1. Divide the data into two groups: F, holes containing only the fixed picks, and C, holes containing only the constraints (limits). These can actually be

carried in the same dataset, with type code (blank or BELOW) indicating to which group a given point belongs.

2. Using only the data in group F, create a grid. A convenient way to restrict the points is to use BELOW codes as control to change the data value to Null temporarily.

3. Interpolate a value from the grid at the locations of each of the holes in group C—that is, at the holes not used to create the grid.

4. Test each of the interpolated values against the limits. This may be done with adequate calculation capability in the data-input portion of the program or with an external program. A constraint is violated if the interpolated value is greater than (i.e., above) a BELOW pick. The constraint is not violated otherwise.

5. Each data point in which a constraint is violated must be transferred from group C to group F. The limiting value is assigned as the fixed value in F. This is easily done by changing the BELOW code to blank.

6. If any holes are added to group F in Step 5 of this iteration, go to Step 2 to create a new grid. If no holes are added to group F (i.e., no limits are violated), stop.

A potential modification to the procedure could occur at Step 5. As written, the observed limiting value is used for a violated hole; this would make the next grid exactly honor that value. However, it might be argued that the surface would be recognized if the drill had just touched it. If so, perhaps we should force the grid deeper than the base of the hole; this could be done by placing the new value to be honored a specific amount (say, 1 foot) below the limit. A greater value could be used to force the surface deeper, though selection of such a value could be difficult.

Application of the procedure is straightforward and, while tedious, should not present difficulties. However, the maximum possible number of points should be used in Step 2. If the first grid goes badly out of control upward in a specific region, all shallow holes in that region will be violated and their TD values used, causing the grid to be a poor representation of the surface locally. In such a situation, we might wish to intervene and use only the deepest constraint, rather than all constraints, in the area. In any event, constraint violations are minimized with evenly distributed data.

Difficulties in the procedure can arise under certain conditions. Whenever a grid is used for mapping, best results are obtained if no grid-squares contain more than one data point. If data are clustered and a given grid-square contains several points, typical mapping packages will tend to honor an average of the data values but none of the actual values. It is therefore possible that constraints will not be honored, even after the procedure is applied, if the nonpenetrating points occur in clusters.

In addition, it is possible for isolated points to continue to violate constraints after application of the procedure. This can be a result of using an interpolation procedure that is not a true inverse of the gridding process, but only approximates a value on the grid. Even though the grid generally represents the surface well, in detail the calculation may be slightly on the wrong side of the limit.

If errors are found after data points have been "corrected," a modification can be made in the procedure; however, it is rarely necessary. At each fixed data point that has been converted from a limit and in which an error has been detected after an iteration through the procedure, do the following: (1) Compute the error—that is, the difference between the limiting value and interpolated value; (2) add or subtract that difference to the well value (depending on the direction of error) to overcompensate; and (3) using that modified data, regrid. If the next iteration still shows an error, correct further.

This procedure is admittedly pragmatic and could cause a fixed point to be poorly honored. However, it generally cures the problem immediately, and in our experience no points ever failed to decrease in error. This problem is generally associated with data-density problems (clustering) or too-coarse grid intervals and is a result of using a discontinuous grid to simulate a continuous variable. Jones and Jordan (1975) made use of this procedure to work with heavily clustered data and obtained good results. The changes necessary to correct the grid are normally small and may not warrant the extra effort.

The procedure for handling limits can be generalized to nonstructural and non-BELOW cases. If thickness is to be mapped, we could calculate it at each hole with a fixed pick. For the nonpenetrating holes, calculate the observed thickness from the horizon to the base of the hole. This would also define a limit (on thickness), but now it specifies that the grid must be greater than this value (that is, it is an ABOVE limit). The modifications of the procedure should be obvious.

When dealing with certain types of surfaces (for instance, the oil-water contact in a reservoir), it is possible for limits to be defined as either ABOVE or BELOW as well as to define ranges in which limits are simultaneously placed on both sides of the grid. This procedure is discussed in chapter 10.

If a shallow horizon is known to be conformable to that being gridded, an alternative approach is to use a variation of the conformable isochore method discussed in chapter 6. In this case the isochores between the two surfaces would be gridded, with missing data giving Null values for thickness. The isochore grid would be modified as needed for violations, and then hung from the conformable horizon above. This may give more realistic results in areas without fixed points.

Authors' Note: Methods are being developed to create grids directly that take such limits into account. Dubrule and Kostov (1986) and Kostov and Dubrule (1986) present theory and methods for calculating a grid value through use of splines and kriging in the presence of inequality constraints.

CHAPTER 6

Building the Stratigraphic Framework

Gridding for single-surface projects was covered in chapter 4; multiple-surface projects involving stratigraphic frameworks are discussed in this chapter. A stratigraphic framework (described in chap. 2) is a group of sequences, each composed of a set of conformable horizons bounded by unconformities. This chapter describes techniques used to construct a set of grids that accurately represent the horizons in a framework. These grids must include such relationships as conformity, truncation, and baselap.

The first section of this chapter describes grid operations, and the remaining four sections discuss the steps involved in building a stratigraphic framework. The first of the framework-building sections describes the use of geologic interpretation in building grids. The second covers techniques used to build grids of surfaces that are conformable to one another. The third section applies to intersecting horizons and describes techniques to incorporate baselapping or truncating relationships into grids. The fourth section discusses how techniques for handling conformity, truncation, and baselap are combined with geologic interpretation to build a correct stratigraphic framework.

Although faults must be incorporated during the grid building process, they are not discussed in this chapter. Chapter 8 describes several methods for putting faults into grids and how those methods relate to the framework building process described here. Faulting has been separated from building stratigraphic frameworks to simplify discussion and to reduce the amount of information presented at one time.

GRID OPERATIONS

Before we look at framework-building procedures, an important computer mapping capability — grid operations — must be discussed. There are two types of grid operations: those that operate on a single grid, and those that operate on two grids (grid-to-grid). Single-grid operations involve a simple calculation (e.g., addition, subtraction, multiplication, division, comparison against minimum or maximum limits, logarithms, square roots, and so on) at each intersection value. For instance, to convert a thickness grid from feet to meters, every grid Z-value is divided by 3.28.

Operations between two grids normally work on a node-by-node basis. Each node in a grid corresponds to an X-Y location. Hence, if two grids cover the same area and have the same grid interval, corresponding nodes or intersections (e.g., row 1, column 1) represent the same X-Y location. We can therefore make calculations using values from matching nodes. For example, subtracting one grid of structural elevation from another to generate a new grid of thickness involves the following steps:

1. Match a given node in one grid with the corresponding node in the second;
2. Determine the Z-value associated with each node;
3. Subtract one Z-value from the other to give thickness; if either node contains a Null, a Null should result;
4. Perform this procedure for every node to create the grid (Fig. 6.1).

Note that this procedure is a computer version of cross-contouring. Some programs allow operations between grids that do not have coincident node locations. Interpolated values from one grid at the other's node locations are used for performing the operation.

A variety of operations other than subtraction can be performed between grids (see Appendix A), but four grid-to-grid operations are most important for construction of a stratigraphic framework: addition, subtraction, and retaining the minimum or maximum value of two grids. The application of each of these operations is discussed in detail in the following sections.

For most of the procedures discussed in this book, the grids used as input to these operations must represent elevation, so node values will be positive above sea level and negative below sea level. If the grids are not in terms of elevation, application of the techniques as described will produce incorrect maps.

GEOLOGIC INTERPRETATION

In all complex mapping projects, two types of information must be supplied. The first and most obvious is data on the surfaces: horizons picks from wells, time

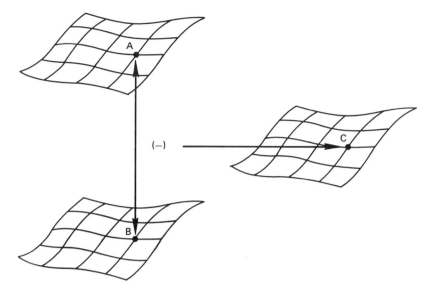

Figure 6.1   Grid-to-grid operations normally work node by node on a grid. Nodes A and B represent the same locations on the top and at the base of a stratigraphic unit. By subtracting B's value from that of A, the thickness of the unit is determined and assigned to node C. A new grid of thickness is created by performing this calculation at every node. This operation is the computer's version of cross-contouring.

to seismic reflectors, topographic survey points, derrick floor elevations, and so on. The second type, as important as the first, is geologic interpretation. Interpretation ranges from correlating picks from well to well or outcrop to outcrop, to working out the sequence of events that led to the present stratigraphic and structural configuration. This interpretive work is done by the project geologist prior to either manual or computer mapping.

The first step in analyzing multiple horizons is to identify the geometric relationships between horizons that are to be mapped. One way to begin is to sketch several "rough" cross sections showing all the important structural and stratigraphic features. Figure 6.2 shows such a cross section through a hypothetical area; this section, discussed below, contains unconformities and baselapping horizons and is used to demonstrate the modeling procedure.

Identifying Unconformities

Unconformities are involved in most of the geometric relationships between geologic surfaces, so recognizing them is crucial. Unconformities are normally known from previous studies in the region or from interpretation of well, outcrop, or seismic data. For someone new to an area, one way to become familiar with

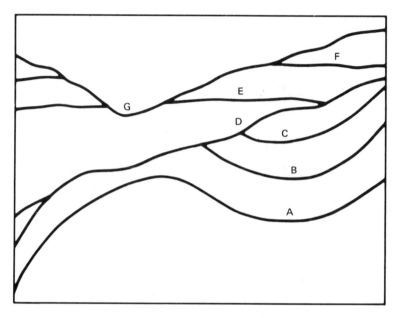

Figure 6.2 Cross section through a hypothetical area showing: (1) two unconformities (D and G), (2) a truncating relationship between D and surfaces A, B, and C, (3) a baselapping relationship between surfaces E and F and surface D, and (4) a truncating relationship between surface G and surfaces E and F. (After Jones and Johnson, 1983, p. 1416)

surface relationships is to use the rough cross sections mentioned above. Choose a section that shows most of the geologic complexity in the area and identifies all horizons that represent unconformities. Angular relationships between surfaces indicate unconformities.

In the example section (Fig. 6.2), surfaces A, B, C, and D demonstrate angular relationships. Surfaces B and C terminate against D. Surface A approaches surface D at a high angle but never actually encounters it, although in another portion of the area it may contact D. Surface D is an unconformity and represents an erosional event that has removed portions of surfaces A, B, and C through truncation.

Surfaces E and F also demonstrate angular relationships; they lap onto surface D from above. Here sediment has been deposited on the unconformity, and surfaces E and F have a baselapping relationship with surface D. Finally, unconformity G has an angular truncating relationship with surfaces E and F.

Figure 6.3 shows the variety of angular and concordant relationships possible between sequences and unconformities. Unconformities generally appear in cross section as continuous surfaces against which other surfaces terminate (Fig. 6.3A). However, some unconformities show an angular relationship only above or below, or not at all (Fig. 6.3B-E). Unconformities that show a concordant relation with

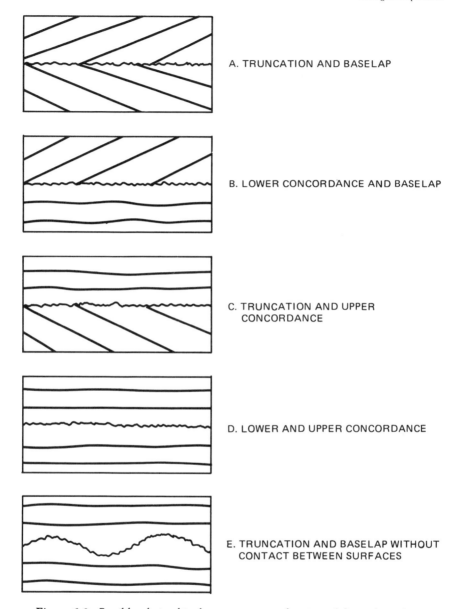

A. TRUNCATION AND BASELAP

B. LOWER CONCORDANCE AND BASELAP

C. TRUNCATION AND UPPER CONCORDANCE

D. LOWER AND UPPER CONCORDANCE

E. TRUNCATION AND BASELAP WITHOUT CONTACT BETWEEN SURFACES

Figure 6.3 Possible relationships between an unconformity and the surfaces above and below it.

a sequence above or below can usually be grouped with that sequence during mapping.

Sometimes an unconformity is concordant to surfaces both below and above it (Fig. 6.3D). This is usually very difficult to recognize in cross section, and an understanding of the local or regional geology is needed to make such an interpretation. Treating the unconformity as conformable to the surfaces above and below it (i.e., as if all surfaces are in the same sequence) will produce an acceptable map. For unconformities that are not concordant to surfaces above or below yet do not intersect those surfaces (Fig. 6.3E), it is usually best to treat the unconformity as having a discordant relationship.

Identifying Sequences

A *sequence* is composed of one or more conformable surfaces that are bounded on the top and base by unconformities. Conformable surfaces generally do not exhibit angular relationships with surfaces other than unconformities. If all unconformities have been correctly identified, intervening surfaces will belong to sequences. In the cross section of Figure 6.2, surfaces E and F are grouped as a sequence bounded above by unconformity G and on the base by unconformity D. Surfaces A, B, and C are conformable and represent another sequence bounded above by unconformity D, but not bounded on the base by an illustrated unconformity. When the lowest surface of interest is part of a sequence, the lack of a lower bounding unconformity is of no concern. Collection of deeper data would show an unconformity at the base.

The identification of most sequences is simple. However, situations exist in which a surface would be identified as an unconformity on the basis of angular relationships, but the geologist groups it as a member of a sequence. In these situations the angular relationship is most likely caused by a unit that has pinched out. Most pinch-outs are low-angle features that are ignored because standard mapping procedures for conformable surfaces (discussed below) work reasonably well. If grids generated by the standard procedures are uncontrolled in the area of the pinch-out, the more sophisticated techniques presented in chapter 9 should be used.

Some pinch-outs, such as those associated with stream channels or reefs (chap. 2), may thin abruptly, causing the two bounding surfaces to intersect at a high angle (Fig. 6.4). When this happens, one or both of the surfaces that bound the thinning unit should be classified for mapping purposes as having either a truncating or a baselapping relationship. These surfaces have the geometry of typical truncation or baselap, even though they do not meet the traditional geologic definition of an unconformity.

In the case of the stream channel (Fig. 6.4A), an erosional event has taken place but was so localized or of such short duration that the geologist has not identified it as an erosional unconformity (Crosby, 1912). Surface C represents the base of the

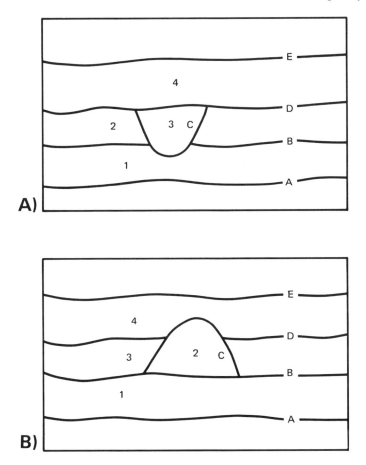

Figure 6.4 Cross section showing angular relationships within a sequence. (*A*) Stream channel (unit 3) is bounded by surfaces C and D. Surface C appears to be truncated by surface D, and surface B is truncated by Surface C. (*B*) Reef (unit 2) is bounded by surfaces C and B. Surface C appears to lap onto surface B, and surface D appears to lap onto surface C.

stream and can be interpreted as truncating surfaces below it. The stream need not encounter the deeper surfaces, but the angular relationship with surface B shows that it does for this surface.

Surface C may be considered a special surface within the sequence; it has a truncating relationship with surfaces below it. Surface D represents both the top of the channel rocks (unit 3) and the top of unit 2 and is a member of that sequence because it is conformable to surfaces A, B, and E. However, surface D "truncates" special surface C and therefore prevents C from incorrectly projecting higher than it should.

99

In Figure 6.4*B*, surface C marks the boundary between the reef material (unit 2) and the adjacent material (units 3 and 4). Surface C does not represent an unconformity and might be grouped with the sequence composed of conformable surfaces A, B, D, and E. However, surface C has an angular relationship with surfaces B and D that is too abrupt to map effectively using conformable-surface techniques. Surface C, which seemingly "laps" onto B, is a special surface within the sequence. Surface D has a baselapping relationship with C and thus terminates against it. Surfaces A, B, D, and E, even though C occurs between them, would be considered conformable members of the same sequence.

The above discussion has presented terms and methods used to describe geometric relationships of surfaces to be mapped, as follows: (1) most projects merely require identifying unconformities and the sequences they bound; (2) conformable surfaces within a sequence may merge due to pinch-out; and (3) if the sequence contains a surface of limited extent that laps onto or truncates surfaces within the sequence, then that surface is a special, nonconformable member of the sequence.

When mapping, the selection of which surfaces to treat as conformable, unconformities, or special depends upon scale or areal extent of the mapping project. For example, over a small area, say less than the width of a river's flood plain, a stream channel may appear to be a major unconformity. Likewise, a regional unconformity may truncate the underlying sequences at such a low angle in this area that they appear conformable. In short, computer mapping allows us to account for geometric relationships between surfaces, though definition of such relationships does not involve hard and fast rules.

CONFORMABLE SURFACES

The second step in building a stratigraphic framework is to generate an initial grid of each surface. For unconformities that do not represent the concordant top or base of a sequence, and for special surfaces within a sequence, this involves gridding the data directly (chap. 4).

For a sequence of conformable surfaces, gridding all surfaces directly will not produce the best results. The fact that surfaces are conformable often provides additional information that can be used to augment the actual data. Problems can arise if conformability of the surfaces is not considered (Walters, 1969; Fontaine, 1985). Fewer wells usually penetrate deep horizons than shallow, causing data density to decrease for deeper surfaces. This variation in density significantly affects the shape of the resulting surfaces, but it can be corrected by using conformable mapping techniques.

For example, the two surfaces in Figure 6.5 have been interpreted as being conformable; however, fewer points exist on the lower of the two surfaces. Simply gridding these surfaces will not produce the structural low in the deeper surface, and the grid (dashed line) may cross the upper grid. Another problem with directly

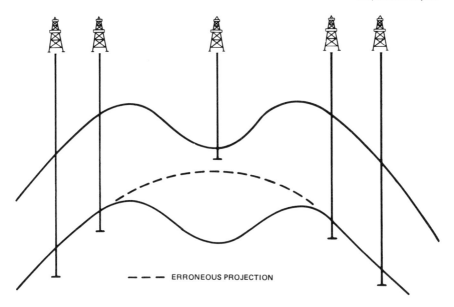

Figure 6.5 Conformable surfaces showing erroneous projection when grids are generated independently. (After Jones and Johnson, 1983, p. 1418)

gridding conformable surfaces having different amounts of control is the variation in complexity of the resulting grids. The grid generated for the upper surface of Figure 6.5 is more complex than the lower grid (dashed line), even though conformable surfaces should have similar forms.

The manual approach commonly used to solve this problem (Handley, 1954; Bishop, 1960; Walters, 1969) includes the following steps:

1. Construct a map of the upper or more densely drilled surface;
2. Calculate thickness at the points common to both surfaces;
3. Use these to draw a thickness map;
4. Subtract the thickness map from the map of the upper surface by cross-contouring, forming a map of the lower surface.

Grids of conformable surfaces are generated by the computer using thickness grids and techniques equivalent to this manual approach (Jones and Johnson, 1983). The process is essentially the same as done by hand:

1. Identify the control surface in a sequence (this is the surface that will build the best grid, normally because it has the most data) and grid it;
2. Calculate the thickness between the control surface and the adjacent conformable surface at the wells penetrating both surfaces;
3. Build an isochore grid using these thickness values;

101

4. Subtract or add (depending on whether the conformable surface is below or above the control surface) the isochore grid from the control grid; use only positive and zero thickness values.

The remainder of this section enlarges on these procedures.

The resulting grid, when contoured, will honor the data where values for both surfaces exist, will be conformable to the control surface, and will contain most of the complexity of the control surface. Gridding isochores rather than directly gridding structure has an additional advantage: Thickness is commonly smoother than structure and therefore has less potential for extrapolation errors.

Isochore grids may extrapolate to negative values away from the data. Negative values in the grid are not incorrect; conformable horizons can intersect (pinch out), and thickness gradients near a pinch-out should cause this projection. Negative grid values are useful for creating subcrop maps and for calculating volumes accurately. Some gridding algorithms do not allow extrapolation below the smallest thickness, preventing negative values. This characteristic of a gridding algorithm may produce a result as incorrect as an algorithm that allows wild extrapolations based on data gradients. Because of differences in how extrapolation is handled, the isochore grids (or the conformable grids generated from them) should be checked to ensure that pinch-outs occur correctly. Chapter 9 discusses techniques for controlling the pinch-out location or zero-line.

Although isochore grids should project to negative values in pinch-out areas, the negative values cannot be used to create conformable grids. For instance, subtraction of an isochore grid containing negatives from a control grid will create a conformable grid that crosses the control rather than being coincident with it through the pinch-out area. Therefore, only positive values should be subtracted from the control grid. Isochore grids containing negative node values and surfaces that "incorrectly" cross are useful for other mapping situations and are discussed in chapters 7-10 and 12.

Two methods involving isochore grids can be used to build conformable surfaces; both are based on the general procedure described above. The first, Method 1, is simple, but requires that the wells be vertical and that wells with picks for a conformable surface also have picks on the control surface. Method 2 is more complex, but can be used for both deviated and vertical wells and for wells without picks on both conformable surfaces. The following two subsections present details on the isochore methods.

Conformable Surfaces—Isochore Method 1

The following steps are required for Isochore Method 1. An example demonstrates how Isochore Method 1 is applied. Figure 6.6 shows an example sequence that contains four conformable surfaces; surface B is designated as the control surface. Processing takes place in the following order.

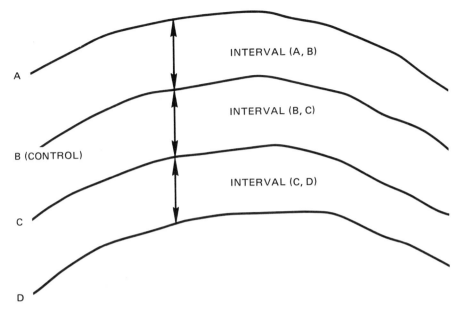

Figure 6.6 Surfaces A, B, C, and D are members of the same sequence, with surface B the control surface. If all wells are vertical and penetrate at least as deep as surface B, Isochore Method 1 can be used to generate structural grids for this sequence.

1. Select and grid the control surface for the sequence. Normally this will be the surface having the most data. However, because of data distribution the control could have fewer points but make a more representative grid. Example: Build a grid of surface B.
2. At each well (X-Y location) compute the thickness for the zone between each pair of adjacent surfaces in the sequence. If the value for either the top or the base of the zone being calculated is missing, no thickness can be calculated and a Null value must result. Example: Calculate the thickness of zones (A,B), (B,C), and (C,D) at each well.
3. Build a thickness grid for each zone. Example: Build grids of thickness for zones (A,B), (B,C), and (C,D).
4. Create the conformable surface below the control by subtracting the thickness grid for the zone immediately below the control surface from the control grid. Example: Subtract thickness grid (B,C) from grid B, creating grid C.
5. The grid generated in Step 4 will now be considered the "new" control surface for generating the next-lower surface. Subtract the thickness grid for the zone below the new control surface from the new control grid generated in the previous step. Example: Subtract thickness grid (C,D) from grid C, creating grid D.

103

6. Repeat Step 5, working downward until grids are constructed for all surfaces within the sequence and below the control surface.

7. Starting over with Step 4, repeat the procedure, working up from the original control surface. In this case, thickness grids are added to control grids to create conformable grids. The result is a grid constructed for each surface within the sequence and above the control surface. Example: Add thickness grid (A,B) to grid B, creating grid A.

The application of Isochore Method 1 is limited to vertical wells in which each well having picks on conformable surfaces also has a pick on the control surface. If these restrictions are not met, conformable grids from Method 1 may not honor all the data. For example, A is the control surface in Figure 6.7. Data for this example are contained in part A of Table 6.1. Zone (A,B) thickness at wells 2 and 4 cannot be calculated because values for A and B do not exist at the same X-Y location (Table 6.1, part B), so the thickness grid is generated with fewer data points.

When the isochore grid is subtracted from grid A to create grid B, surface B points at wells 2 and 4 will not have been used for construction of grid B, so grid B may not exactly honor those values. In Step 5, the next-lower conformable surface is used as a temporary control surface to construct the conformable surface below

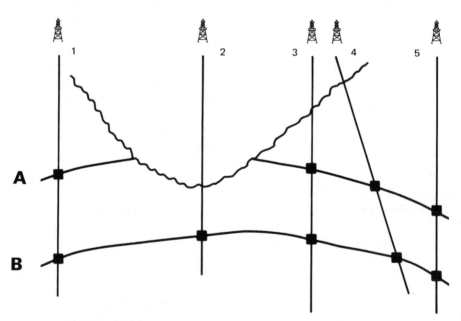

Figure 6.7 Two conformable surfaces, with surface A the control surface. Wells 1, 3, and 5 are vertical and penetrate both surfaces. The control surface is eroded at well 2, and well 4 is deviated. If Isochore Method 1 is used, picks for surface B at wells 2 and 4 are not used to generate grid B.

it, so several adjacent conformable surfaces have varying data distributions. This problem is solved by using Isochore Method 2.

Conformable Surfaces—Isochore Method 2

The following steps are required for Isochore Method 2. Figure 6.8 shows four conformable surfaces, with surface B designated as the control surface. Processing takes place in the following order.

1. Select and grid the control surface for the sequence, using the same criteria as with method 1. Example: Build a grid of surface B.
2. Interpolate from the control grid the structural elevation at each data point (X-Y location). Example: Interpolate an elevation at each data point from grid B.
3. In the data file, replace all Null values in the field of the control surface by the interpolated values. Do not change any of the original data values. Example: Replace all Null values for surface B with interpolated values.
4. At each well (X-Y location), compute the thickness for the zone between the control surface and the conformable surface to be gridded. Because missing values on the control surface have been replaced by interpolated values, a calculated thickness exists for every real value on the conformable surface. Example: Calculate thicknesses for zones (A,B) and (B,C) at the wells.
5. Build a thickness grid for that zone. Example: Build grids for zones (A,B) and (B,C)
6. If the conformable surface is below the control surface, subtract the thickness grid for the zone immediately below the control surface from the control grid. If conformable surfaces lie above the control surface, thickness grids are added to control grids to create conformable grids. Example: Subtract thickness grid (B,C) from grid B, creating grid C; add thickness grid (A,B) to grid B, creating grid A.

TABLE 6.1
Example Data for Section of Figure 6.7

		Part A			Part B
X	Y	ID	A	B	(A,B)
5639	18750	Well 1	−3004	−3057	53
5757	18710	Well 2	Null	−3038	Null
5896	18760	Well 3	−2996	−3044	48
5883	18751	Well 4	−3012	Null	Null
5901	18749	Well 4	Null	−3070	Null
5950	18743	Well 5	−3049	−3098	49

105

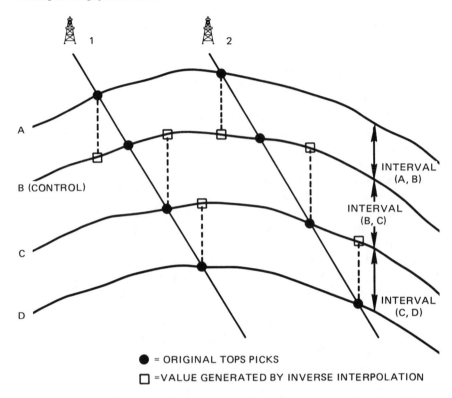

● = ORIGINAL TOPS PICKS

☐ =VALUE GENERATED BY INVERSE INTERPOLATION

Figure 6.8 Surfaces A, B, C, and D are members of a sequence, with surface B the control surface. Wells 1 and 2 are deviated. Isochore Method 2 should be used to generate structural grids for this sequence.

7. Each grid generated in Step 6 in turn will be considered the "new" control surface for generating the next-lower and -higher surface. Repeat Steps 2 through 6 until all conformable surfaces for the sequence have been generated. Example: Interpolate an elevation from grid C at each data point; in the data file, replace all Null values for surface C with interpolated values; calculate thicknesses for zone (C,D) at the wells; build a grid of thickness for zone (C,D); subtract thickness grid (C,D) from grid C, creating grid D.

Conformable Data-Merge

The control surface should be well controlled if conformable grids are to be built from it, but in some cases the control surface may not produce an acceptable grid. For example, in Figure 6.9 control surface A has been removed by erosion in the area where well 2 penetrates it. The eroded portion of A contained critical

106

information about the structural low in the sequence at that location. Gridding surface A directly will produce a grid having the shape of the dotted line; the grid should more correctly follow the shape of the dashed line.

Surface B contains information about the structural low that could be of assistance in building grid A. The conformable techniques described above are used to build a better grid A than surface A data alone would build. This is done by temporarily using surface B as control while building grid A and then reversing the process to build final grid B and all others conformable to A. The authors commonly use a modified Data-merge procedure (Conformable Data-merge), similar to that described in chapter 5, to enhance surface A data with data from surface B.

Besides the difficulties that may arise if a good grid cannot be built for the control surface, problems with methods 1 and 2 may occur if the areal data distribution for conformable surfaces differs. For example, Figure 6.10 shows a sequence of dipping conformable surfaces that have been truncated. Because of this truncation, the areal distribution of data varies from surface to surface. Picks for lower surfaces are found only on the left side of the section, with more picks to the right for higher surfaces. Few points are vertically above each other, so Method 1 does not work and Method 2 has problems because the areal distribution of points on any one surface is restricted.

Conformable Data-merge can use data from surfaces above and below to enhance those surfaces with insufficient data. Once the data for these surfaces have been enhanced, Isochore Method 1 or 2 can be used to build the final grids.

Recall from chapter 5 that the Data-merge procedure involves the following steps.

Build a grid for one dataset.

Interpolate values for point locations in a second dataset from that grid.

Calculate the difference between original and interpolated values for each location in the second dataset.

Build a residual grid using the calculated differences and user-supplied values at specified ring points.

Shift values in dataset(s) by adding or subtracting the difference interpolated from the residual grid.

If the datasets in the Data-merge procedure represent horizon picks from two conformable surfaces, the residual grid is equivalent to the thickness between those surfaces, so shifting values in one dataset by the residual grid amount is equivalent to adding or subtracting the thickness grid from the control-surface grid. Interpolation from the initial grid is equivalent to using inverse interpolation with deviated well data in Isochore Method 2.

There are two differences between Data-merge and Isochore Method 2: (1) ring points allow more control of the residual (isochore) grid in Data-merge and (2) the

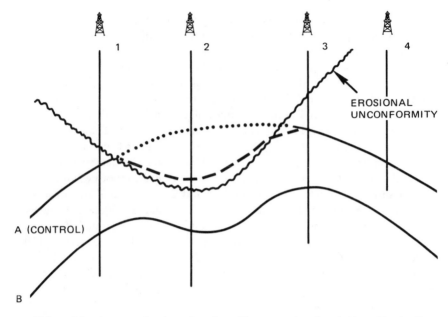

Figure 6.9 An example where directly gridding control surface A (dotted line) will not produce the "best" grid. Use of Isochore Method 2 and surface B as a temporary control surface will produce a more correct surface (dashed line). Conformable Datamerge could also be used to enhance surface A data.

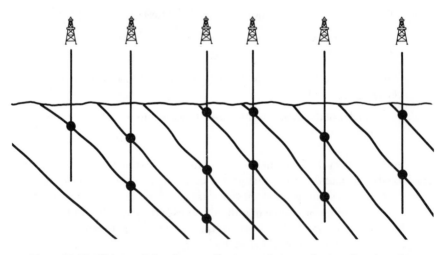

Figure 6.10 If the areal distribution of horizon picks varies from surface to surface, Isochore Method 1 cannot be used and Method 2 may have problems.

108

result from Data-merge is a combined set of data, rather than a final grid. We must therefore make a choice: Should ring points be added to Isochore Method 2, or should the Data-merge procedure be modified to work with conformable surfaces? Either of these approaches provides a useful mapping tool. We have chosen to describe and demonstrate a modified Data-merge procedure because enhancing the data, rather than producing a final grid, leads to a larger family of applications.

The following modifications to the Data-merge procedure convert it to a Conformable Data-merge procedure:

1. One dataset rather than two is used; it contains a separate field for each surface;
2. The grid interval for the residual (isochore) grid is based on the data distribution of the residual values, not a coarse grid interval designed to produce a smooth residual;
3. The values assigned to the ring points are not zeros, but are reasonable estimates of the thickness between the conformable surfaces.

A simple problem involving two conformable surfaces with varying amounts of data demonstrates the Conformable Data-merge procedure. From interpreted seismic data, a shallow, good reflection is identified at shot-points on nine lines (Fig. 6.11A); a deeper reflection is poorly developed and was interpreted only on portions of seven lines (Fig. 6.11B). The upper (control) surface data produce an acceptable grid and map; however, the lower surface can be gridded over only a small portion of the map area. Knowing that the two surfaces are conformable, we can use the shape of the upper surface to augment the data of the lower surface in the southeast portion of the map area.

Conformable Data-merge for this example requires the following steps.

1. Call the control surface data field A; this is the field in the file that will make the best grid. Designate field B as the field containing the data to be augmented. Grid the data in field A.

2. Calculate the difference (thickness) between original values of the field to be augmented (field B) and values interpolated from the control surface (field A) grid at the B locations. Do this by subtracting the interpolated values of field A from the original values of field B. The original values of field A could be used; however, the error between interpolated and real values is usually small and unimportant when building the isochore, so no effort is made to combine real and interpolated values.

3. Build an isochore grid using the thickness values and the ring points described in the Data-merge procedure. Use a grid interval appropriate for the data distribution of the thickness values. Assign values to ring points that are representative of the thickness between the two surfaces near each ring point. Use a negative thickness if the surface being augmented (field B) is deeper than the control surface (field A).

4. Use the isochore grid to shift the control-surface (field A) data points to the surface being augmented. Using inverse interpolation, calculate values at all data

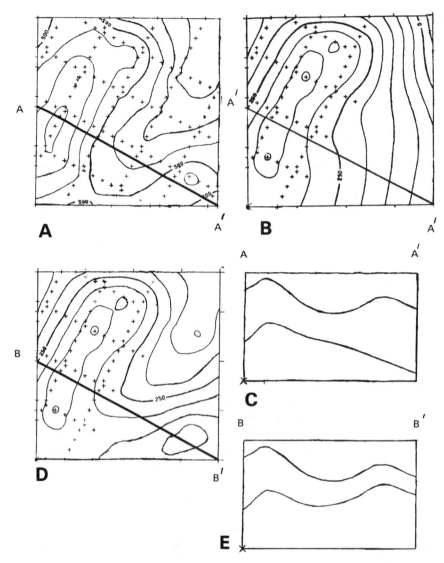

Figure 6.11 Maps and cross sections showing use of Conformable Data-merge for augmenting data on a conformable surface having restricted data distribution. *(A)* Upper control horizon. *(B)* Lower horizon. *(C)* Cross section showing grids if Conformable Data-merge is not used. *(D)* Lower horizon after augmentation with Conformable Data-merge. *(E)* Cross section showing grids after use of Conformable Data-merge.

locations from the residual grid. Add the interpolated values to the original control-surface data values (field A). Replace all Null values in the field of the surface to be augmented (field B) by the calculated values, but do not change any original values of field B.

The data field for the lower surface will now contain the original data for that surface, plus data augmented from the upper surface where previously there were Nulls. Gridding and contouring the modified lower-surface data produces the map in Figure 6.11D. Notice that the surface is similar to the upper horizon in the east but honors the original data in the west. The cross sections in Figure 6.11C and E show the upper and lower surfaces before and after using Conformable Data-merge.

Enhancement Using Several Surfaces

Conformable Data-merge is especially appropriate in situations where several surfaces are used to assist in defining the shape of one surface. Often it is necessary to construct a single structure map for a stratigraphic surface that, although present, is not recognizable everywhere. Surfaces lying conformably above and below the surface to be mapped can aid in defining its shape. In this case several surfaces may be needed because no single surface covers all the areas where data are missing from the surface to be mapped.

A similar problem also arises in regional studies, not because of incomplete data, but because stratigraphic units in one area are defined differently or are undefined in other areas. A stratigraphic horizon at the top of a prominent formation at one location may be defined as a marker bed within a formation in a neighboring area, and in still another area, although present, it may have been considered so minor that it was not recorded. To map this horizon over the entire area would first require recorrelation to obtain picks where they were not recorded and to recheck previous picks for proper correlation. A reasonable and much faster approach, although not always as precise, uses conformable surfaces above and below to assist in defining missing portions of the surface.

Figure 6.12 shows a cross section containing three conformable surfaces. For various reasons, picks have been recorded for only small portions of each surface. The goal is to map all three surfaces over the entire area. Gridding each surface independently may produce acceptable results; however, it is more likely to produce grids that extrapolate incorrectly in areas where data were not collected. Data for each surface can be enhanced with picks from the other two surfaces by use of Conformable Data-merge.

The general approach is to use data from the best surface to enhance the surfaces above and below. The enhanced surfaces are then used to enhance other surfaces. Once the highest and lowest conformable surfaces have been encountered, the process is repeated in the opposite direction — that is, the surface whose data was last enhanced is now the "best" surface, and surfaces above and below it are enhanced. This chain of enhancements is repeated until the surface that

111

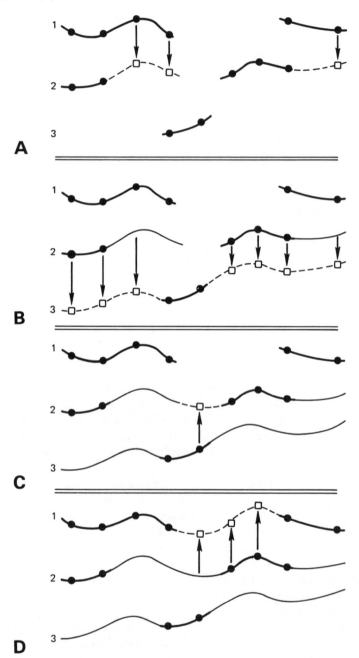

started the process has been modified. Using this approach, all conformable surfaces within a sequence potentially assist in the building of other surfaces within that sequence.

In the example shown in Figure 6.12, Conformable Data-merge must be performed four times. The four merges are as follows:

1. Because surface 1 will produce the best grid, it is used as the A data field. Surface 2 data, used as the B-field, are enhanced in the areas where surface 1 is present and surface 2 is missing. Ring points are assigned thickness values representative of the (A,B) interval near them. Note that surface 2's data are enhanced only where surface 1 has data and that none of surface 2's original data are modified (Fig. 6.12A).

2. The enhanced surface 2 data are used as the A-field and surface 3 data as the B-field, to augment the surface 3 data (Fig. 6.12B).

3. The enhanced surface 3 data are used as the A-field and surface 2 data as the B-field (Fig. 6.12C).

4. The most recently enhanced surface 2 data are used as the A-field and surface 1 data as the B-field (Fig. 6.12D).

The merges described above are needed only if the A data field contains information not contained in the B data field.

Authors' Note: This section describes isochore methods for building grids of conformable surfaces. These methods work well if the dips on the surfaces are not steep and variable. As discussed in the structure section of chapter 2, isochores of intervals that are folded with steeply dipping flanks will be significantly thicker on the flanks than in the crests and troughs. An isopach grid will show less variation across the fold and therefore be more accurate than an isochore grid.

Jim Downing and Robert Duncan (1986, personal communication) point out that, with a special program, isopach thickness can be created from a grid of the control surface and the picks for the conformable surface. The isopach values are then gridded. The grid nodes of the control surface must then be shifted by the isopach thickness in a direction normal to the control surface. This shifted data and the original data for the conformable surface are used to build the structure grid of the conformable surface. Presently, most mapping programs do not have the ability to make these isopach calculations. However, few projects have structural complexities that warrant this extra effort. Further, the isopach and isochore methods will give similar results for dense data distributions.

Figure 6.12 Cross section showing three conformable surfaces with data recorded for only small portions of each surface. Conformable Data-merge allows portions of each surface to be used in augmenting data for the other surfaces because portions are filled progressively in Steps A-D. Arrows represent the estimation of values from conformable surfaces.

113

TRUNCATION AND BASELAP

The third step in building a stratigraphic framework is to incorporate truncations and baselaps into the grids. When a grid is built, no geologic information
about the relation between that surface and the surfaces above and below it is used,
so the grids in a sequence can cross those in another sequence. Figure 6.13 shows
the initial grids for the stratigraphic framework described at the beginning of this
chapter. Rather than correctly terminating at unconformitites as in Figure 6.2, the
grids project past one another.

The three possible relationships between a surface and those below it are

1. conformable — the surface is quasi-parallel to the surface below and, if it
   contacts the lower surface, it does so at a low angle,
2. truncation — the surface truncates the surfaces below it, and
3. baselap — the surface laps onto an unconformity or special surface below it.

A fourth nongeologic relationship — "first" — is included for completeness. The
lowermost surface for which data were collected has an unknown relationship to

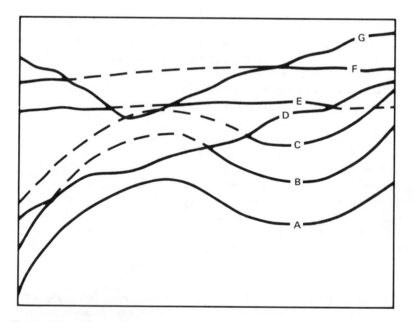

Figure 6.13 The initial grids of a stratigraphic framework are built independently
and will therefore cross one another. Truncations and baselaps are built into these
grids using grid-to-grid operations. (After Jones and Johnson, 1983, p. 1416)

114

the surfaces below it, and since it is the first or oldest surface, it will be assigned a "first" designation.

Truncation

Figure 6.14A shows grids A and B, which, because they were built independently, cross one another. Because surface A has been interpreted as a truncator,

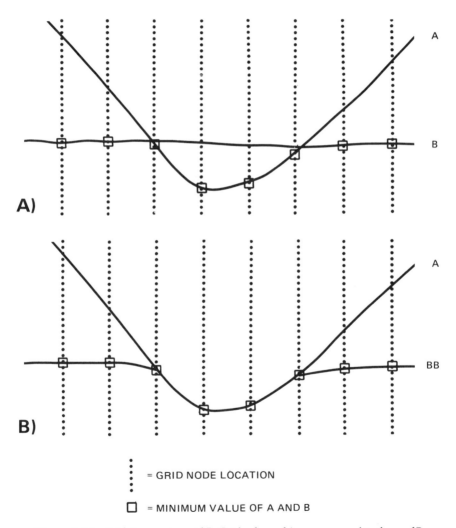

A)

B)

⋮ = GRID NODE LOCATION

☐ = MINIMUM VALUE OF A AND B

Figure 6.14   Grid A truncates grid B. Grid values of A are compared to those of B and the minimum elevation is output, creating a new truncated grid BB.

115

surface B must be prevented from incorrectly projecting above surface A. To truncate one grid by another, the elevation Z-values at each grid node are compared, and the minimum Z-value is retained in a new truncated grid (Jones and Johnson, 1983). Figure 6.14B shows grid A and newly created grid BB. These grids properly portray the truncating relationship between the two surfaces. In the area of truncation, grids A and BB have the same Z-values and so are coincident.

A truncating surface can potentially affect every surface below it. The section in Figure 6.13 shows surface D truncating surfaces C and B but not contacting surface A. However, because surface D may cut surface A in another area, the general rule is: Every grid below a truncating surface should be treated as if it is intersected by the truncator. If a lower grid does not actually contact the truncating grid, the "truncated" grid will be identical to its pretruncation equivalent.

Baselap

Figure 6.15A shows initial grids for surfaces A and B. Surface A has been interpreted as lapping onto surface B, so grid A must be prevented from incorrectly projecting below B. To lap one grid onto another, the elevation Z-values at each grid node are compared, and the maximum value is retained in a new grid (Jones and Johnson, 1983). Figure 6.15B shows newly created grid AA and grid B. These properly portray the baselap relationship between the two surfaces. In the area of baselap, grids AA and B are coincident.

Every surface within a sequence that has a baselapping relationship to the unconformity below it can potentially lap onto that unconformity. Even if a surface within a sequence does not show baselap in a given cross section, the surface may contact the unconformity elsewhere. As with truncation, a "lapped" grid that does not intersect the unconformity will be identical to its nonlapped equivalent. Only surfaces in the sequence immediately above an unconformity can have a baselapping relationship to that unconformity, so the baselap relationship ends at the next-higher unconformity.

Truncation and baselap techniques prevent grids from crossing. However, they do not precisely locate the point of intersection. In Figure 6.15B, the lap of grid AA onto grid B shifts slightly at the contact on the left. This shift occurs because grids (a set of distinct points) are used to represent continuous surfaces. The grids can be precisely controlled only at the nodes, so unless a node occurs at the point of truncation or baselap, the contact between the two surfaces will not be exactly located. This problem will always occur with grids but can be minimized by reducing the grid interval.

For most situations, these truncation and baselap techniques are acceptable. However, they are not used for calculating volumes when high precision is required. In those situations, crossing (pretruncation or prebaselap) grids are used (chap. 10). Crossing grids are also used for subcrop displays (chap. 7).

116

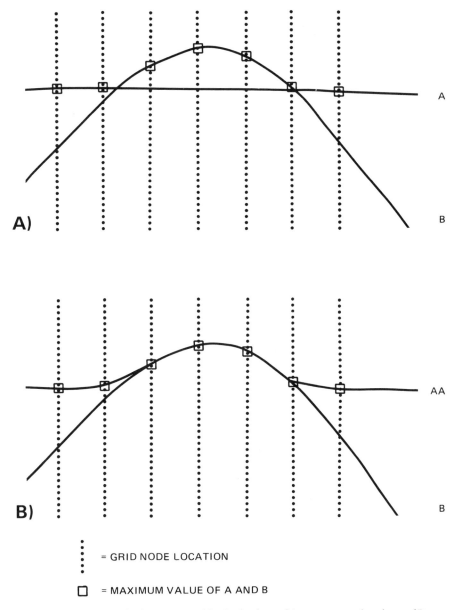

= GRID NODE LOCATION

☐ = MAXIMUM VALUE OF A AND B

Figure 6.15 Grid A laps onto grid B. Grid values of A are compared to those of B and the maximum elevation is output, creating a new grid AA that properly laps onto B.

117

COMBINING INTERPRETATION, CONFORMITY, TRUNCATION, AND BASELAP

The process of building a stratigraphic framework is complex for anything but the simplest geologic interpretation. The complexity is not caused by the variety of different procedures that must be performed, for there are only four: grid creation, isochore calculation, truncation, and baselap. The complexity results from the number of times and the order in which these procedures must be performed. Leaving out a step such as a truncation, or changing the order in which steps are performed, can totally alter the geologic relationships. Fortunately, some basic rules for setting up these procedures simplify the process.

A recipe for specifying instructions to the mapping program in building a stratigraphic framework centers around the framework table; the interpretation going into the table and the mapping instructions are implied by table entries (Figure 6.16). The *framework table* provides a means for ordering the conformity, baselap, and truncation relationships to account properly for the geologic interpretation. A simple set of rules then guides the conversion from framework table to program instructions, allowing correct stratigraphic framework grids to be built.

Creating the Framework Table

This section defines the framework table and its various entries. The stratigraphic cross section shown in Figure 6.2 is used as an example; its geologic interpretation was discussed previously. Table 6.2 shows a general form of the framework table. Each of the seven columns contains a summary of the allowable entries for that column. The following discussion explains the entries in each column (italicized paragraph headings) and demonstrates the generation of the framework table (Table 6.3) corresponding to the stratigraphic framework of Figure 6.2.

Surface name. In this column, enter the code name or number used to identify each surface during the mapping process, one per line. Order the code names from the topmost surface downward (youngest on top and oldest on the bottom), just as they appear in the geologic column.

Surface type. Three possible surface types exist: unconformity (UNCF), surfaces within a sequence (SEQ), and special surfaces (SPEC). A surface is referred to as an unconformity when it has a discordant (truncation or baselap) relationship to

Figure 6.16 Flow chart for building a stratigraphic framework. The framework table orders the geologic relationships to simplify the specification of mapping program instructions.

118

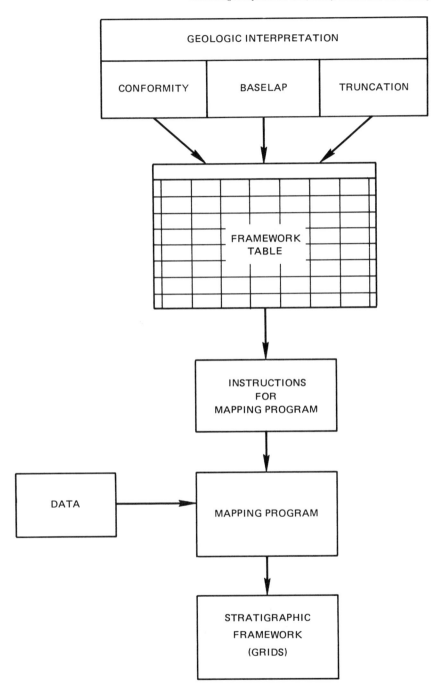

TABLE 6.2
Allowable Entries in the Framework Table

Surface name	Surface type	Sequence number	Sequence control	Initial grid-building method	Primary baselap/ truncate	Special baselap/ truncate
Names or IDs (young)	Uncon-formity (UNCF), Sequence (SEQ), or Special (SPEC)	If UNCF, X	If UNCF, X	If UNCF, DIRECT	FIRST, BASELAP, or TRUNCATE	X, BASELAP, or TRUNCATE
		If SEQ, every surface in sequence assigned same number (1 for oldest, 2 for next, etc.)	If SEQ, CONTROL or CONFORM If SPEC, X	If CONTROL, DIRECT If CONFORM, ISOCHORE If SPEC, DIRECT	If SPEC, X	
(old)		If SPEC, sequence number containing surface				

surfaces above and below it. All conformable surfaces within a sequence, plus concordant bounding unconformities, are of the sequence type. A special surface is a nonconformable member of a sequence; it has a limited areal extent relative to the map area and an intersecting relationship with one or more surfaces in the sequence.

Sequence number. The sequence number provides a means for designating which surfaces belong to the same sequence. Starting from the bottom and working upward, all surfaces belonging to the first (oldest) sequence are assigned 1, all surfaces in the second sequence are assigned 2, and so on. Special surfaces are assigned the number of the sequence within which they are found. Unconformities are not given a number; X is used instead.

Sequence control. One surface within a sequence must be designated as a control surface (CONTROL), and all others within that sequence are designated as conformable to that control (CONFORM). Usually the control will be the surface having the most data. However, because of data distribution the control surface could have fewer points but make a more representative grid. Special surfaces are nonconformable members of a sequence, so this terminology does not apply and X

120

TABLE 6.3
Framework Table for Figure 6.2

Surface name	Surface type	Sequence number	Sequence control	Initial grid-building method	Primary baselap/ truncate	Special baselap/ truncate
G	UNCF	X	X	DIRECT	TRUNCATE	X
F	SEQ	2	CONTROL	DIRECT	BASELAP	X
E	SEQ	2	CONFORM	ISOCHORE	BASELAP	X
D	UNCF	X	X	DIRECT	TRUNCATE	X
C	SEQ	1	CONFORM	ISOCHORE	FIRST	X
B	SEQ	1	CONFORM	ISOCHORE	FIRST	X
A	SEQ	1	CONTROL	DIRECT	FIRST	X

is used. X is also used for unconformities (surface type = UNCF) because they are not members of any sequence.

Initial grid construction method. One of two methods is normally used to construct the initial grids: direct gridding or the conformable isochore procedure. Unconformities, special surfaces, and control surfaces are normally built by direct gridding (DIRECT), although supplemental points or isochore techniques must sometimes be used to control grid construction. One of the two conformable isochore methods is used to build grids of conformable surfaces (ISOCHORE).

Primary baselaps and truncations. Intersecting relationships include truncation and baselap. Truncation implies that this surface or sequence potentially truncates all surfaces below it. Truncation is usually applied to unconformities. However, if the lowest surface in a sequence is an unconformity grouped with the sequence because of a concordant relationship, then truncation is applied to the sequence, and grid-to-grid operations for truncation are performed between the lowest grid of the sequence (the unconformity) and all lower grids. Baselap implies that grids in the sequence lap onto the underlying unconformity or onto the upper surface of the underlying sequence if it is a concordant unconformity. An additional entry in this column—FIRST—designates the lowest surface or sequence and that no grid-to-grid operations are to be performed. The same designation—FIRST, BASELAP, or TRUNCATE—applies to all surfaces of a sequence. Special surfaces are not included in the primary set of baselaps and truncations, and their positions should be marked X.

Special baselaps and truncations. A second series of baselaps and truncations is required to incorporate special surfaces into the framework. Rules similar to those for the primary case hold for these cases, although they need be implemented only for surfaces that could potentially intersect the special surfaces. All surfaces below the lowest special surface are assigned an X. The special surface is assigned the appropriate term, BASELAP or TRUNCATE. If the special surface is assigned TRUNCATE, then the surface above it is also assigned TRUNCATE. If the

121

special surface is assigned BASELAP, all surfaces of that sequence above the special surface are assigned BASELAP, because they could potentially lap onto the special surface. If no other special surfaces exist, all remaining surfaces are assigned X's. If another special surface exists, the remaining surfaces below it are assigned X's, and the designation process is repeated.

Converting the Framework Table to Detailed Steps

The framework table now contains all information required to build the stratigraphic framework. The rules for converting the table information to the detailed steps of framework construction are as follows.

1. Build a grid for each surface in which direct grid construction is appropriate.
2. For all conformable surfaces having the same sequence numbers, use the appropriate isochore method and the control surface for that sequence to build conformable grids.
3. Starting at the bottom (oldest surfaces), perform the appropriate baselap or truncation grid-to-grid operations. No special surfaces should be included in these operations.
4. Starting at the bottom, perform the appropriate baselap or truncation grid-to-grid operations on the special surfaces and all surfaces that they affect.

Applying these rules to the framework table (Table 6.3), we perform the following steps.

1. Build initial grids for surfaces requiring direct grid construction.
     Build grid A.
     Build grid D.
     Build grid F.
     Build grid G.
2. Build initial grids for conformable surfaces in sequences 1 and 2. Use either Isochore Method 1 or Method 2 as described above. In sequence 1, surface A is the control, and thicknesses are added to build surfaces B and C. In sequence 2, surface F is the control, and thickness is subtracted to build surface E.
3. Perform the primary baselap and truncation grid-to-grid operations.
     Truncating relationships for surface D
     Compare grid D to grid C and output the mininum (deepest) elevation, creating grid CC.
     Compare grid D to grid B and output the minimum elevation, creating grid BB.
     Compare grid D to grid A and output the minimum elevation, creating grid AA.

Baselapping relationships for sequence 2

Compare grid E to grid D and output the maximum (shallowest) elevation, creating grid EE.

Compare grid F to grid D and output the maximum elevation, creating grid FF.

Truncating relationships for surface G

Compare grid G to grid FF and output the minimum elevation, creating grid FFF.

Compare grid G to grid EE and output the minimum elevation, creating grid EEE.

Compare grid G to grid D and output the minimum elevation, creating grid DD.

Compare grid G to grid CC and output the minimum elevation, creating grid CCC.

Compare grid G to grid BB and output the minimum elevation, creating grid BBB.

Compare grid G to grid AA and output the minimum elevation, creating grid AAA.

4. Perform the special baselap and truncation grid-to-grid operations.

There are no special surfaces.

The final grids for this stratigraphic framework are AAA, BBB, CCC, DD, EEE, FFF, and G.

This example does not demonstrate the application of the framework table with special surfaces. A modification of the stratigraphic framework (Fig. 6.17) shows the framework of Figure 6.2 with the addition of a small stream channel (surface S) that has cut through surface B. Table 6.4 shows the framework table for this example.

Application of the rules for expanding this framework table generates all of the steps of the previous discussion, plus appropriate steps for incorporating the special surface. Following are the additional required steps.

1. Build a grid for each surface.

Build grid S.

4. Perform the special baselap and truncation grid-to-grid operations.

Truncating relationships for special surface S

Compare grid S to grid BBB and output the minimum elevation, creating grid BBBB.

Compare grid S to grid AAA and output the minimum elevation, creating grid AAAA.

Truncation for portion of sequence 1 above special surface S

Compare grid CCC to grid S and output the minimum elevation, creating grid SS.

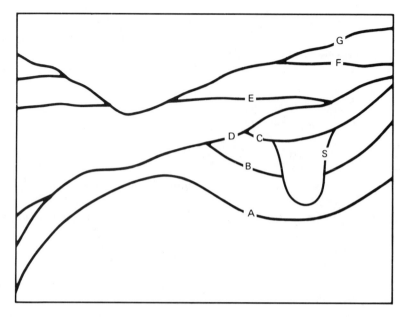

Figure 6.17 The stratigraphic framework of Figure 6.2, with the addition of stream channel surface S. Surface S is a special surface within the sequence composed of surfaces A, B, and C. Surface S truncates surface B and appears to be truncated by surface C. (After Jones and Johnson, 1983, p. 1416)

TABLE 6.4
Framework Table for Figure 6.17

Surface name	Surface type	Sequence number	Sequence control	Initial grid-building method	Primary baselap/ truncate	Special baselap/ truncate
G	UNCF	X	X	DIRECT	TRUNCATE	X
F	SEQ	2	CONTROL	DIRECT	BASELAP	X
E	SEQ	2	CONFORM	ISOCHORE	BASELAP	X
D	UNCF	X	X	DIRECT	TRUNCATE	X
C	SEQ	1	CONFORM	ISOCHORE	FIRST	TRUNCATE
S	SPEC	1	X	DIRECT	X	TRUNCATE
B	SEQ	1	CONFORM	ISOCHORE	FIRST	X
A	SEQ	1	CONTROL	DIRECT	FIRST	X

The final grids for this stratigraphic framework are AAAA, BBBB, SS, CCC, DD, EEE, FFF, and G.

The framework table may at first appear confusing or cumbersome. However, its use will force consideration of all the stratigraphic relationships and their incorporation into the grids. Whether or not the table is filled out formally, all these considerations are needed to build the grids.

Although the framework table is not used directly in other portions of this book, the surface relationships defined in the table are used throughout. The table is a helpful reference when ordering the steps to build subcrop maps (chap. 7), to calculate volumes (chap. 10), and to restore paleogeology (chap. 12).

CHAPTER 7

Displaying Stratigraphic Relationships

Chapter 4 discusses creation of a grid, and chapter 6 extends the concepts to multiple horizons in a stratigraphic framework. A single contour map is not adequate to depict complex stratigraphic relationships, and simply generating a structural contour map for every grid in the framework is not appropriate. For example, portions of a surface that do not exist because of truncation should be left blank on the map. Mapped subcrop lines indicate the intersection of two structural surfaces and outline limits of erosion or nondeposition.

A contour map with subcrop lines shows structural configuration and complex relationships over the map area, while a cross section shows the relationships of surfaces and horizons along a profile. This tool allows detailed study of several horizons at once rather than only one, although from a different perspective than do maps. Cross sections and contour maps are both important aids for interpretation and display of stratigraphic relationships.

CROSS SECTIONS

Cross sections and profiles are widely used for geologic analysis and interpretation. They allow detailed study of individual horizons or of such variables as porosity or interval thickness because they show variation along the line of the section. Cross sections are also valuable for illustrating multiple horizons and the stratigraphic relationships between them.

In addition to the primary purpose of geologic analysis, cross sections have two

127

other important applications. The first allows us to determine if the grids under consideration were made correctly, as discussed in chapter 4 (Fig. 4.8). The second application allows a check on the geologic interpretation. Two horizons might have been assigned a truncating relationship, but the cross section might show that baselap is more reasonable. Again, such relationships are more easily detected and interpreted with cross sections than with contour maps.

The cross section is drawn by the computer in a manner similar to hand construction: (1) Select the line for the section; (2) construct the plot area, with distance plotted along the horizontal axis and elevation on the vertical axis; (3) specify grids to be plotted; (4) select grid nodes that fall along or near the line of section; and (5) plot the node values on the display, connecting the points. Several horizons may be plotted on a single set of axes.

Figure 7.1 shows a computer-generated structural cross section through the stratigraphic framework discussed in Appendix B (Table B.1 and Fig. B.1). The grids were built using the data in Tables B.4 and B.5. The section extends from well W5 (northwest) to W10 to W17 (east-central) and shows both truncating and baselapping relationships along the unconformities, indicating where and how the horizons intersect. This pictorial representation provides valuable information on structural similarities and differences, thickness changes, and stratigraphic relationships.

These continuous grids overplot in such a way that truncated horizons merge into the unconformity. Some programs use curve-fitting algorithms to draw smooth lines for grid profiles rather than straight lines between nodes. This process may

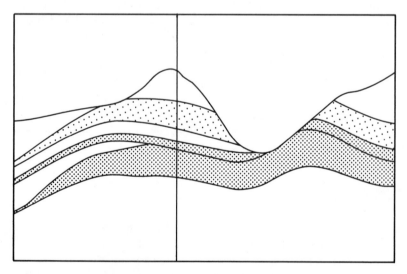

Figure 7.1 Cross section drawn through the stratigraphic framework outlined in Table B.1, based on the data from Table B.4. The line of section is drawn from well W5 to W10 to W17. The stippled areas represent the Aurora Creek and Lamont sandstones.

allow the profile of a sharply truncated or baselapped grid to project past an adjacent grid when in fact the grids do not cross. Some programs optionally allow a straight-line or curve-fitting approach to be used.

The grids in Figure 7.1 were made following the procedures described in chapter 6. Recall that intersecting relationships are entered by operations between grids. For instance, the portion of a surface that is missing because of truncation is made coincident with the truncating surface, rather than stopping abruptly as in nature. These continuous grids overplot in such a way that truncated horizons merge into the unconformity.

Plotting nonstructural grids on a profile shows how the corresponding variables are related. For instance, a plot showing structural elevation, thickness, and porosity allows three important features of petroleum deposits to be studied together. A structural high may be present, but if the reservoir interval at that location is thin or has low porosity, the potential trap may not hold enough oil or gas to be economic. To make such plots, the program must plot and operate with several vertical scales.

Considerations for Building Cross Sections

Several decisions must be made whenever a cross section is to be generated. They include such factors as the location of the line of section, the horizontal and vertical scales, whether the section is referenced to a stratigraphic datum, and projection of data onto the section. These and other considerations are discussed by Bishop (1960), Levorsen (1967), and Langstaff and Morrill (1981).

The first factor — location of the line of section — is basic. The orientation of the section should be chosen on the basis of the features of interest and the nature of the geologic relationships. Recall that apparent dip observed in an outcrop may be less than true dip if the face of the outcrop is not oriented parallel to dip direction. Similarly, only a section that is drawn parallel to dip will show the true dip relations.

A section is commonly drawn from one data location (e.g., a well or outcrop) to another. If other wells exist between the two ends, the section may be located to pass through them, as with the wells in Figure 7.1, establishing a zig-zag pattern. Cross sections thus consist of a series of short segments that are tied together. More consistent results are usually obtained with straighter sections, although substantial changes in local dip direction could lead to each segment's being assigned a different orientation.

A single cross section is usually not adequate to portray the subsurface completely. Several sections should be constructed parallel to the direction of major interest, normally parallel to the dip direction. In addition, at least some crossing strike sections should be drawn. Contour maps are usually necessary to give the complete geographic picture.

A cross section or profile must be assigned a horizontal scale, much like the scale on a map. In fact, using the map scale simplifies referencing between map and

129

section. The vertical scale is assigned separately and ideally should be the same as the horizontal scale. However, vertical variation in a structural horizon is generally low compared to the length of a cross section, and the vertical and horizontal scales are necessarily different.

Use of different scales introduces vertical exaggeration, causing distortion in dipping relationships (Langstaff and Morrill, 1981). If two horizons dip at similar, low angles, their differences will be accentuated. Vertical exaggeration reduces apparent differences if the horizons dip steeply. Although cross sections ideally should not have vertical exaggeration, it is generally necessary to expand the vertical dimension to provide sufficient detail.

Observed well data are important and should be put on the cross section. Posting the picks at wells along the line of section gives specific information about local control. The picks also indicate whether the grid honors general trends in the data or is forced to bend sharply to honor individual points.

If the section does not pass through a number of wells, picks are commonly projected onto the line of section. To do so, a distance limit is established, and wells within this limit are posted. Some programs use the grid to aid in this projection, while others project horizontally. Be cautious, as slopes off the line of section often influence how accurately the projected points match the grid at the section.

Stratigraphic Sections

Figure 7.1 is a structural section, and the plot is referenced to the sea-level datum (elevation). This type of section shows structural features such as dipping beds, folds, and faults. A stratigraphic section uses a horizon as the reference datum. The selected horizon is flattened, and the other horizons are plotted above or below that datum according to the interval thickness relative to that horizon. This is particularly useful if thickness variations are obscured by extreme structural relief. As part of the process of removing structural complexity, stratigraphic sections also clarify relations between unconformities and other horizons.

Some programs calculate a datumed section directly, requiring only specification of which horizon is the datum. In such a case the datum horizon is flattened, and the other horizons are adjusted to be comparable. Alternatively, grid calculations are used to create referenced grids for use in plotting the section. For example, three horizons are represented by grids A, B, and C in Figure 7.2A. To plot a stratigraphic section with A as the datum, create three new grids by subtraction of reference A from each of the others:

$$AA = A - A$$
$$BB = B - A$$
$$CC = C - A$$

130

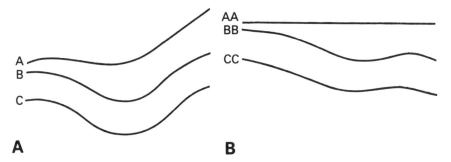

Figure 7.2 *(A)* Structural cross section showing horizons A, B, and C. *(B)* Stratigraphic section showing same horizons, labeled AA, BB, and CC, as referenced to a datum of horizon A.

Note that all Z-values of AA equal zero, and BB and CC have negative or zero values. Plotting AA, BB, and CC from an "elevation" of zero gives the stratigraphic cross section.

Best results in plotting stratigraphic cross sections are obtained if the reference horizon does not intersect the other horizons or is not an unconformity. This reference horizon should not reflect paleotopography. Further discussion of cross sections, including procedures to account for restored truncation, is found in chapter 12.

SUBCROP MAPS

If a horizon has not been truncated, does not lap onto a deeper horizon, or does not pinch out, the continuous grid derived from the procedures described in chapter 6 is appropriate for drawing a contour map. This structure grid does not have intersecting relationships and so would be contoured according to the principles discussed in chapter 4.

However, if a given horizon has been truncated in some portion of the area under study, its contour map should be blank in the truncated region. The horizon is absent there and hence should not be depicted. If the contours were extended onto the erosional surface where the horizon is truncated, the resulting map would depict a misleading hybrid of stratigraphic and unconformity surfaces. Similar considerations apply to baselapping relations. The continuous framework grids described in chapter 6 represent such mixtures of two or more geologic surfaces.

Contours on a structural surface should stop at sequence boundaries (unconformities) or pinch-outs. However, the continuous grids contain only partial information about the location of these boundaries and should not be used for drawing contour maps. If one grid is the unconformity bounding a sequence and another is a stratigraphic surface in the sequence, then their intersection represents the sequence boundary or subcrop line. When the stratigraphic surface is contoured, grid nodes on one side of the line are valid, while those on the other side are not and should be blanked out. Creation of a subcrop map involves plotting subcrop lines and blanking out missing regions; this requires additional grid manipulations and special contouring instructions.

First consider the case of truncation and the subcrop line in the structure map of Figure 7.3, built using the data presented in Appendix B (Table B.4). These contours show the structure on the top of the upper Lamont surface. The blanked region indicates where the surface was removed by truncation (unconformity UNCF) and is marked by the single bounding subcrop line. Creation of this map is not a difficult task.

Suppose horizon B is truncated by unconformity A. The first step in creating the framework (chap. 6) is to generate simple nontruncated grids. Figure 7.4A shows grid B intersecting grid A before the grids are operated against each other. The next step is to perform the truncation operation described in chapter 6 to create a continuous grid. However, for subcrop maps this operation is modified to create a discontinuous grid, BB (Figure 7.4B). As before, output Z-values of B to BB at those nodes where B is deeper than A, but put Nulls in BB at those nodes where B is shallower than A.

This operation is the second dealing with truncation; the operation in chapter 6 creates continuous grids, but this procedure creates a discontinuous (i.e., contains Nulls at the truncated portion) grid that is suitable for subcrop mapping. If the grid is contoured, the map will be blank in the region containing Nulls and will have structural elevation where B is present, showing the structural configuration of horizon B. (Note: The grid should be used only for this purpose; errors will result if this discontinuous grid is used to build up other portions of a stratigraphic framework or to calculate volumes.)

Contours show the form of the horizon, and blank areas show the truncated portion. However, determination of the exact edge of truncation is difficult, particularly in the dip direction, because the map is blank past the last contour; an edge must be inferred from the contours and gradient. Addition of the subcrop line clarifies the erosional edge. The contours and subcrop line can be obtained with the following steps.

1. Build the discontinuous grid BB by operating pretruncation grid B against A; output Nulls for nodes where B is above A.
2. Contour grid BB; the map will be blank where BB contains Nulls.
3. Using pretruncation grids, create grid $I = A - B$ by subtraction. This grid will have positive Z-values where horizon B exists and negative values where the horizon is missing — that is, where grid B projects shallower than grid A. The boundary between the positive and negative node values represents the edge of truncation.

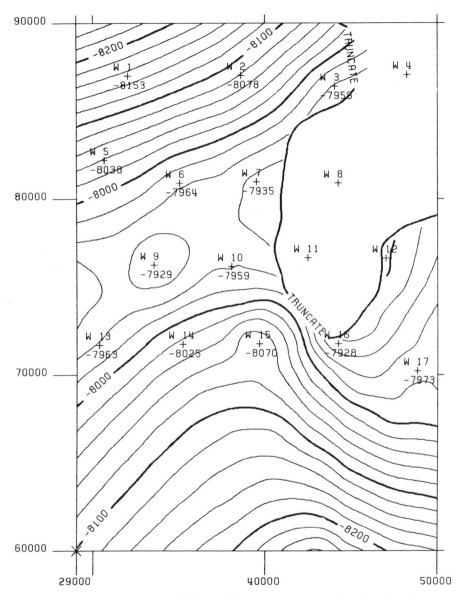

Figure 7.3 Structure map of the top of the upper Lamont surface, based on data from Table B.4. The blank portion corresponds to the nonexistence of the horizon due to truncation by unconformity UNCF. The subcrop line delineates the edge of erosion.

133

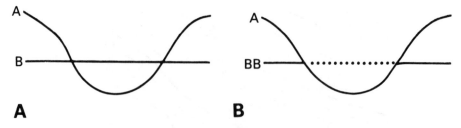

Figure 7.4 Cross section showing truncation of horizon B by horizon A. *(A)* Grids represent the configuration before truncation operation. *(B)* Section after discontinuous grid BB has been formed by truncation operation that sets "eroded" nodes to Null (represented by dots).

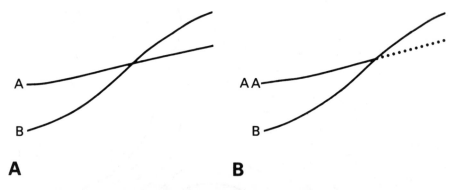

Figure 7.5 Cross section showing horizon A lapping onto horizon B. *(A)* Grids represent the configuration before baselap operation. *(B)* Section after discontinuous grid BB has been formed by baselap operation that sets "nondeposited" nodes to Null (represented by dots).

4. Overplot grid I on the previously drawn map of BB. Only the zero contour is plotted, and this single line marks the subcrop line. The structural contours may not extend all the way to this line because plotting algorithms do not normally project into Nulled regions, and the line is within partially Nulled grid-squares.

A similar process is used for baselapping relationships. Assume horizon A laps onto horizon B with the configuration shown in Figure 7.5A. Only the portion of A on the left exists and should be mapped. The process of generating a structure map with subcrops is straightforward. It requires only two grid operations and two contour plotting steps: (1) Create discontinuous grid AA, containing Null values where the horizon is absent as grid A projects deeper than grid B (Figure 7.5B); (2)

134

plot the discontinuous grid, giving structural contours where the surface exists and blanks elsewhere; (3) create the difference grid $I = A - B$; and (4) overplot the zero contour from grid I on the structure map.

However, the task will be more complex if a younger unconformity lies above the two grids discussed. In this case, the grid previously calculated for the surface being mapped (either AA or BB) is used for all operations. The unconformity grid must be operated against the stratigraphic grid calculated previously (BB in the first example and AA in the second), resulting in additional Nulls where the grid is above this unconformity.

To generate the subcrop line, subtract the stratigraphic grid BB calculated previously from the unconformity, giving a grid with positive, negative, and Null nodes. The Nulls represent the portion of the grid that was affected by the previous baselap or truncation; no subcrop lines will be drawn through that area. This process can be repeated as many times as the surface was affected by later truncations.

A pinch-out represents nondeposition within a conformable sequence. The process of drawing the pinch-out line and blanking the contours is basically the same as that used for baselap and truncation, although it differs slightly because conformable surfaces are built by adding or subtracting an isochore grid from a structure grid. Again the process requires only two grid operations and two contour-plotting steps.

1. Generate an isochore grid for the unit below the surface being mapped. Since the unit is part of a sequence — that is, the top and basal surfaces are conformable — this grid normally will already have been created when building the framework (chap. 6). This grid should be positive where the unit exists and negative where it has pinched out (chap. 9).

2. Create a discontinuous grid for the structural top that contains Null values where the isochore grid is negative.

3. Plot the discontinuous structure grid, drawing structural contours where the surface exists and blanks elsewhere.

4. Overplot the zero-contour from the isochore grid on the structure map.

In many cases, the conformable sequence within which the pinch-out is found has a baselapping relationship with a lower unconformity. Here, the subcrop procedure for baselaps will have been performed and a modified structure grid containing Null values created. If this modified structure grid exists, then the isochore grid must be modified before it is used in Step 2. To do this, create a new isochore grid that contains Nulls where the structure grid has Nulls due to baselap.

Modifying the structure grids to account for all baselaps, truncations, and pinch-outs and to draw subcrop lines can involve many steps. These steps are simple for any single subcrop line, although combining them may seem complex. The task is simplified if the order of processing moves from oldest to youngest.

135

ISOCHORE MAPS

An isochore map shows the vertical thickness distribution of a specified mapping interval. These maps are used to illustrate the size and shape of local basins or depositional centers, either after deposition ceased or after erosion. Further, paleotopography can be interpreted through variation in either the thickness of the eroded interval below or the thickness of material deposited over the unconformity (Maslyn and Phillips, 1984). Isochore maps are also used to help define history of growth faulting, to interpret broad structural movement, and to determine structural relationships responsible for a given type of sedimentation. These maps have a number of other uses and are discussed in more detail by Lee (1955), Bishop (1960), Levorsen (1967), and Paynter (1970), among others.

Again, the term *isopach* represents contours of the stratigraphic thickness of a unit rather than vertical thickness (isochore). Differences between the two are minor if bedding dips at less than 20-30 degrees. However, true thickness differs from vertical thickness in more steeply dipping beds. If true stratigraphic thickness is important, corrections should be made to the thicknesses that are used by the program (cf. Badgley, 1959; Savoy and Valentine, 1961; Pennebaker, 1972). However, the grids that would be obtained are not appropriate for constructing a stratigraphic framework (chap. 6) or computing volumes (chap. 10).

Mapping isochores is not particularly difficult, but special techniques are necessary. Several different procedures apply, depending on the interval being mapped. The basic cases include: (1) a conformable interval within a sequence, with neither the top nor the base of the interval intersecting another horizon; (2) an interval between an unconformity and a stratigraphic surface in the adjoining sequence (with either truncating or baselapping relationships); (3) an intra-sequence conformable interval that is cut by an unconformity or that laps onto an unconformity; and (4) an interval that includes two or more sequences.

The first case is simplest. Here the interval has conformable horizons on the top and base, and neither of these horizons intersects another horizon (although they might intersect one another). Assume the framework has been constructed according to the methods described in chapter 6. If the two horizons defining the interval are adjacent, Isochore Method 1 or 2 (discussed in chap. 6) makes the grid directly; this grid would then be contoured.

Structure grids made with the Isochore Method are built by progressively using thickness between the various units. All available information (e.g., slanted wells, more complex structure of other surfaces, and so on) is therefore used in their construction. If other horizons are included within the interval so the isochore grid has not been made, simply subtract the elevation of the base grid from that of the top and contour the resultant thickness grid.

If the interval does not pinch out, then simple contouring as discussed in chapter 4 is adequate. However, if the interval goes to zero thickness, additional control may be necessary to give a good zero-line, as described in chapter 9. In any event, the grid should contain negative Z-values in the region where the interval is absent. Only the zero and positive contours are drawn; the area containing negative nodes is left blank.

The other cases of isochores are illustrated by Figures 7.4-7.8. These indicate the different intervals of interest, defined by horizons A and B. In some situations, an unconformity U is also included. Other framework horizons within the interval are not shown.

The second case deals with a truncation-defined interval (Fig. 7.4A). To contour the thickness between horizons A and B, first calculate thickness grid $I = A - B$ using pretruncated grids. Grid I will have positive values where the interval is present and negatives elsewhere; map only the zero and positive contours. If truncated, coincident grids are used for the subtraction, the Z-values will all equal zero in the missing region, and a poor zero-line will result (chap. 9). The same considerations apply to baselap (Fig. 7.5).

The third case is illustrated in Figure 7.6A, with horizons A and B both truncated by unconformity U. When the initial pretruncated grids were made, they had the form shown in Figure 7.6B — that is, grids AA and BB cross grid U. We use these untruncated grids to generate an appropriate isochore grid.

A convenient procedure for making an isochore grid is to form an envelope — that is, to create a top and a base — and then subtract the basal elevations from the top. This concept is discussed in more detail in chapter 10. Here, create a grid AAA that consists of the lesser (deeper) elevation of the U and AA grids. Grid AAA thus follows grid AA where horizon A exists and follows grid U where horizon A does not exist (Fig. 7.6C). Grid AAA forms the top of the envelope, and grid BB can be used as the base. Subtracting grid BB from grid AAA gives a new grid. This grid contains the thickness between the stratigraphic horizons where A and B both are

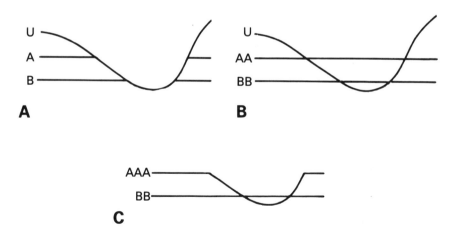

Figure 7.6 (A) Cross section showing horizons A and B truncated by unconformity U. (B) Original pretruncated grids AA and BB cross U. (C) Grid AAA forms the top of the interval of interest, consisting of a combination of U and AA. Grid BB forms the base of the interval. The thickness grid is made by AAA − BB.

137

present and the thickness left after erosion where A is missing; negative values are found where B is missing.

This grid is mapped normally for zero and positive contours. The resulting map will show a well defined zero-line and a change in gradient at the wedge where the unconformity cuts the mapped interval. The contours will be more closely spaced in the eroded portion than in the undisturbed part. If the exact location of the intersection between unconformity U and horizon A is not of interest, the map is complete. However, if this line is important, calculate a grid as $U - AA$ and overplot its zero-contour on the preceding map.

Figure 7.7A shows a related situation. Here horizons A and B lap onto the unconformity. The initial prebaselapped grids AA and BB project through grid U (Fig. 7.7B). These grids are used to form the isochore map. The base of the envelope is defined by the greater (shallower) elevation of the U and BB grids (BBB in Fig. 7.7C). Subtracting this combined base grid from the top of the envelope AA gives the grid to be mapped. The contours and subcrop lines are drawn as in the truncation case.

The fourth case, pictured in Figure 7.8, can be more complex. In Figure 7.8A, the interval of interest covers two sequences, with an unconformity between horizons A and B. If the horizons do not intersect, this situation is handled as with previous cases; subtract grid B from grid A and contour the resulting grid.

Figure 7.8B shows more complexity, as can occur if the situation of Figure 7.8A is further truncated. Again, an envelope must be constructed, using concepts already discussed. However, the complications may be so severe that an exact map is not worth constructing. If so, postbaselap and -truncation continuous grids can be used, but the zero-line will not be drawn correctly (see chap. 9). In such a case, plot only positive contours, and plot a special contour near zero (say, 0.1) to approximate the zero-line.

The concepts that we have discussed regarding isochore mapping also have application in computation of volumes. Chapter 10 points out that volumes are normally calculated by first creating a thickness grid using envelope concepts and then numerically integrating this grid for values above zero. The grids that are used for mapping thickness as discussed above are used for the numerical integrations.

Although thickness maps are valuable tools for studying depositional patterns, they do have limitations in structural analysis. Lee (1955) suggests that structural analysis can best be done if the top and base of the interval were originally planar and horizontal. If one or both of the surfaces is erosional with high relief, the structural trends may be detected only broadly.

Lee (1955) points out that thickness maps may not indicate sequence of events or time of folding because conformity, baselap, or truncation can produce the same thicknesses. He also gives examples of isochore maps that might be valueless or misleading, depending on the definition of the mapped interval. These difficulties are commonly associated with an interval that contains two or more sequences— that is, an unconformity is located between the top and base horizons, as shown in Figure 7.8.

138

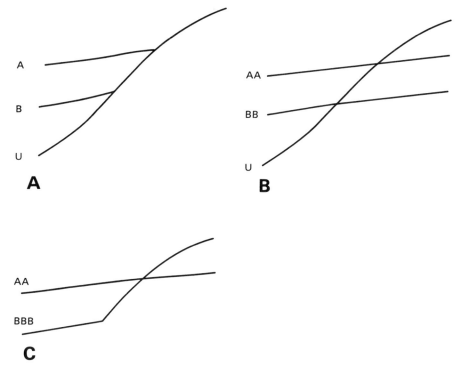

Figure 7.7 *(A)* Cross section showing horizons A and B lapping onto unconformity U. *(B)* Original prelapped grids AA and BB cross U. *(C)* Grid BBB forms the base of the interval of interest, consisting of a combination of U and BB. Grid AA forms the top of the interval. The thickness grid is made by AA − BBB.

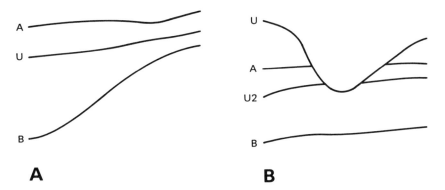

Figure 7.8 Cross section showing the interval of interest (defined by horizons A and B), which contains an unconformity. *(A)* Unconformity is wholly contained, so subtraction, A − B, gives the thickness grid to be contoured. *(B)* Interval cut by two unconformities, U and U2.

CHAPTER 8

Faulting

The ability to incorporate faults is an important aspect of generating contour maps. Some faults are easy to map, while others are extremely difficult (Dickinson, 1954; Hintze, 1971; Fontaine, 1985). In either case, special techniques are usually required to incorporate the faults because most mapping programs assume a continuous surface. If a fault is ignored and mapping techniques described in previous chapters are used to contour across a fault, the contours will continue from one side of the fault to the other. Depending on the nearness of data to the fault and the displacement on the fault, the surface may appear as a steep monocline or a barely perceptible change in slope.

Few, if any, programs automatically identify faults based on input data. Most require that the presence and location of faults be interpreted prior to mapping. Numerous computer methods are available for mapping projects, each with specific advantages and appropriate for particular types of faulting. We describe four methods — fault-block, fault-trace, restored-surface, and fault-plane — in this chapter; all require that fault locations be known. Which method to use depends on available data and assumptions that can be made.

The fault-block method uses polygons to isolate individual fault blocks. Typically, several sides of a fault-block polygon — although not all — are fault traces. Data points or hand-drawn contours are used to estimate grid nodes inside the polygon. Defining these polygons is tedious, and to reduce time and effort, the same set of polygons is often used for several surfaces, therefore treating the faults as vertical. If a dipping fault plane must be modeled, a different set of polygons is required for

141

each surface. This method, though manpower intensive, is appropriate when working with few or with hundreds of faults.

The fault-trace method uses fault-trace locations and a special gridding algorithm to build a continuous faulted grid. This algorithm prevents data on one side of a fault from being used to calculate grid nodes on the other side. As with the fault-block method, the same set of fault traces can be used for several surfaces, implying a vertical fault. However, fault traces are relatively simple to digitize, and shifting the trace for each surface provides an acceptable method of modeling dipping fault surfaces. This method is appropriate when working with few or hundreds of faults.

The restored-surface method uses fault trace locations, throw, and the portion of the surface affected by the fault to build a grid representing vertical displacement. This grid is used to restore horizon picks affected by the fault to their prefault position. The restored picks are gridded, and the grid is then faulted by adding the displacement grid. The restored-surface method restores only the vertical component of fault displacement and is used primarily when faults can be assumed vertical. The method is not practical for more than a few faults or when a new set of traces is required for each surface in the situation of a dipping fault plane.

The fault-plane method is used to map surfaces cut by dipping faults when enough data are available to build grids of the fault planes. Separate grids are built for the surface on each side of a fault; thus, if one surface is cut by a number of faults, several grids represent the surface piecewise. The fault-plane grids are operated against the surface grids to prevent them from projecting across the faults. Because grids are used to define the fault planes, this method is useful only for dipping faults, is not practical for more than a few faults, and can handle reverse faults. The following sections describe each of these methods in detail.

FAULT-BLOCK METHOD

The fault-block, or polygon, method was one of the first methods used to contour faulted surfaces with the computer. The technique is slow, labor intensive, and primitive when compared to sophisticated mapping techniques available today. However, it is still a commonly used technique; it produces an accurate map and is particularly useful for mapping highly faulted areas.

The fault-block method works best with digitized structural contours but will work with other types of data (Walters, 1969). When digitized contours are used, this method provides better control of contours near faults than do the other methods described in this chapter. Chapter 9 discusses why hand-drawn maps are gridded and how the data are collected. Typically, the structural contours are drawn by the geologist and contain all available information about the surface and the faults that cut it. In addition to the hand-drawn contours, original horizon picks from wells and outcrops can also be used as input data. However, the fault-block method usually does not work well with horizon picks alone.

Fault-block mapping is a four-step process (Szumilas, 1977): (1) Divide the map area into fault blocks or polygons; (2) digitize the polygons (fault traces), their associated hand-drawn contours, and horizon picks; (3) build grids for each fault block; and (4) overplot contours from all fault blocks on the same map, with contours restricted to be inside fault-block polygons.

As an example, Figure 8.1A shows a structural horizon cut by four faults. From seismic data the geologist has interpreted the structure and drawn the contours. The goal is to use the computer to reproduce the hand-drawn map as accurately as possible.

Defining the Fault-Block Polygons

Fault-block polygons are used to isolate portions of the surface. Polygons consist of a series of connected straight-line segments (see chap. 4). Fault-block polygons

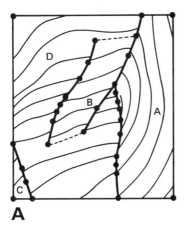

A

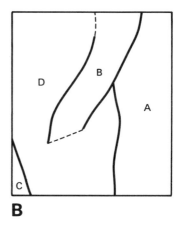

B

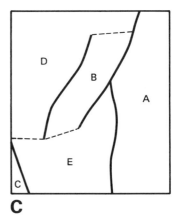

C

Figure 8.1 (A) Contours of a structural surface cut by four faults. Zero-throw fault traces separate the area into four fault blocks. (B, C) Alternative configurations for the fault-block polygons.

143

are bounded by fault traces, the edge of the map, and imaginary lines drawn by the mapmaker. Since a fault trace is a smoothly curving line, several short straight-line segments are used to define it.

The imaginary lines are needed to close polygons involving faults that die out or end within the map area — that is, the faults do not intersect other faults or the edge of the map. These lines are treated as if they were faults having no displacement and so are handled as zero-throw fault traces. Typically, these are single line segments connecting the end of one fault trace to another or to the edge of the map.

Fault blocks should be defined in a manner that produces the fewest and largest polygons, but a particular fault configuration does not lead to a unique set of polygons. The example surface has been subdivided into four polygons (Fig. 8.1A), but it could have been subdivided in other ways (Fig. 8.1B and C). Note that five polygons have been identified in Figure 8.1C. In this case, the mapmaker may have felt that polygon D of Figure 8.1A should be separated into two polygons because polygon B projects into polygon D. However, the characters of the surfaces on either side of polygon B appear similar, and so points on one side of polygon B can be used to calculate grid nodes on the other side. This is a judgment call, and there are no rules for making this kind of decision.

Note that all vertices of the polygons are marked on the maps (dots in Fig. 8.1A). Since a fault segment usually defines the edge of two adjacent polygons, the same points must be used to define that fault for both polygons. Using different sets of points for the same fault in two polygons will result in portions of those polygons overlapping in some areas and being separated by gaps in other areas. Even when attempting to digitize the same point twice, once for each fault-block polygon that borders the fault, it is almost impossible to digitize the same location exactly. Therefore, often the polygon vertices must be edited or a program written to find nearby points on different polygons and make them coincident.

Digitizing the Data

Three types of information may be digitized: polygons, hand-drawn contours, and horizon picks, although often horizon picks will already be in digital form. Digitizing polygons involves recording the X-Y coordinates and a polygon identifier for every vertex of the polygon. The fault-trace indicator is an optional code that can be recorded along with each polygon vertex. It can be used in plotting to separate true fault traces from map edges and zero-throw faults. Record a value of "0" for each point that is on the map edge or is part of a zero-throw-fault trace. Record a "1" for all other vertices of the polygon. Use of this indicator code is discussed in the section on contouring.

Digitizing contours (discussed in chap. 9) involves recording the X-Y coordinates for points along each contour, the value of the contour, and an identifier of the polygon in which that contour belongs. The map must be prepared for digitization; often the original map must be redrawn in order to extend contours manually a short distance beyond the edge of the fault block. This extension ensures that grid nodes accurately represent the surface near the polygon boundary. Extending contours also allows grid nodes outside the polygon to be assigned values related to the surface inside the polygon. When contours or volumes are then calculated inside the polygon, outside nodes allow calculations to be carried more accurately to the edge of the block.

The zero-throw-fault trace is an imaginary line that cuts across a nonfaulted area of the map; contours on one side of the line must connect smoothly with those on the other, even though different grids and polygons are used in their construction. Extending and digitizing contour lines beyond the edge of each polygon ensures that contours drawn to zero-throw-fault traces will accurately connect with the same contour on the opposite side of the zero-throw-fault trace.

Figure 8.2A, B, and C shows the separate fault-block maps constructed for the example. Nonadjacent fault blocks A and C were placed on the same map since they were sufficiently separated that their extended contours and grid nodes did not intersect. On all of these maps the contours extending beyond the fault-block edges reflect the character of the surface, not the surfaces bounding the fault block.

Building the Grids

A grid is built for each fault-block map using all of the contour data associated with that map. The grid interval should be appropriate for the detail at which the contour lines were digitized. Chapters 4 and 9 discuss choosing a grid interval and considerations related to digitized contour data. A given fault block may cover only a small amount of the map area, and calculating values for all surrounding grid nodes can be time consuming and is unnecessary. The gridding process is typically constrained to an area near the fault block; values are assigned to nodes inside and near the limits of the fault block and Nulls to those nodes farther away.

Most projects involve more than one surface, and faults need to be built for each surface. If the faults can be assumed to be vertical through all surfaces, the same set of fault-block polygons is used for each surface. Further, if there is no growth on the faults — that is, interval thickness on opposite sides of the faults is the same — then only one faulted structural surface need be built. To generate the next-lower conformable surface, subtract the isochore grid for the interval below the surface from all fault-block grids for that surface. Repeat the process for each lower surface, and add isochores for higher surfaces.

If the fault planes dip in such a way that they cannot be assumed to be vertical, or if growth is associated with the faults, each structural surface must be constructed independently; in the case of slanted faults, a new set of fault traces must be used for each surface. Repeating the fault-block method for each surface is time consuming, and the advantages of doing so need to be carefully considered.

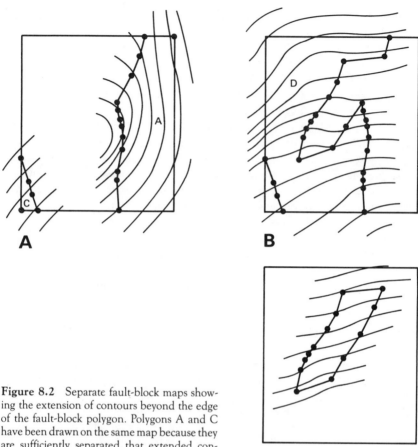

Figure 8.2 Separate fault-block maps showing the extension of contours beyond the edge of the fault-block polygon. Polygons A and C have been drawn on the same map because they are sufficiently separated that extended contours do not interfere.

Applying the Fault-Block Method

Contour Maps

A contour map with faults is generated for the entire surface by overplotting contours for each fault block on the same map; this is done by restricting the contouring to inside those corresponding polygons. However, grid nodes both inside and outside the polygons are used to draw the contours up to the polygon boundary. By overplotting each set of contours over the previous set, a complete contour map is drawn.

Normally the geologist will want the faults displayed on the contour map. The polygons cannot be used to do this because they contain lines related to the

146

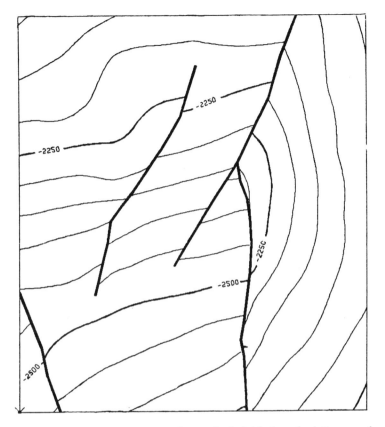

Figure 8.3 Contour map constructed using the fault-block method. Contours from separate fault-block grids were overplotted, with fault traces overplotted on the contour map.

zero-throw-fault traces. If a fault-trace indicator code was used when the polygons were digitized, the appropriate boundaries of the polygons are converted to a set of fault trace lines.

The approach recommended in the digitizing portion of this section is to record an indicator code of "0" for each point on the map edge or part of a zero-throw-fault trace and a "1" for all other vertices along the fault. Each adjacent pair of polygon vertices, including the first and last, can define a segment. Each segment is defined by the coordinates at its start and end; if the trace-indicator code was assigned to each end, line segments for which both codes are zero are deleted. The only remaining lines are those belonging to true fault traces. These lines are then overplotted on the contour map to create the finished faulted contour map (Fig. 8.3).

147

Cross Sections

Cross sections through structural surfaces are important tools for checking the correctness of the grids, as well as for adding to the understanding of structural and stratigraphic relationships. If all the grids that were built for one surface with the fault-block method are displayed together in cross section, it is difficult to interpret the display because the grids project past the fault blocks to which they belong. If the cross-section program has the ability to draw the surface only within the fault blocks and to plot the polygons as vertical lines (faults), then a good cross section can be generated.

An alternative approach is to build a single faulted grid for the entire surface. This is done by combining into one grid only those nodes from each fault-block grid that are inside its fault-block polygon; the grid thereby contains no projected nodes. When all grids have been combined, one grid will exist for each structural surface. These combined grids can be used for cross-section display. They also produce interesting maps because the contours are continuous from one block to the next and give a strong visual indication of the relative throw on each fault.

For normal, nonvertical faulting, the combined grid for a surface will contain Nulls where the structural surface is missing because of the gap associated with the fault face. Values can be calculated for those Nulls by using nearby nodes as data. As a result, nodes in the gap will have values intermediate between node values on the upthrown and downthrown sides of the fault. In this situation, the intersection of the fault plane and the surface on either side of the fault is not angular but is rounded. The number of surface nodes used to calculate the gap values influences the degree of rounding of the surface intersections.

Volumes

The method used and assumptions made during the construction of faulted grids can have a significant effect on calculated volumes. The assumption of vertical faulting and the method used to simulate surfaces can cause grids to differ from true structure. The difference between the grids and the true surfaces must be understood if we are to have confidence in the generated volumes and recognize what is lost or gained by the grid construction method. We discuss four possible methods.

Method 1. Figure 8.4A and B shows a structural contour map and cross section. The cross section depicts the true position of the material between the mapped surface and a lower surface. If fault traces on the upper surface are used to define the fault-block polygons, dashed lines in Figure 8.4C indicate where the polygons bounding the two sides of the fault intersect the lower surface. The polygon has the effect of forming a boundary that passes vertically from the upper surface to the lower surface.

When fault block grids and polygons are used to calculate volumes between the two structural surfaces, two major differences between true and calculated vol-

148

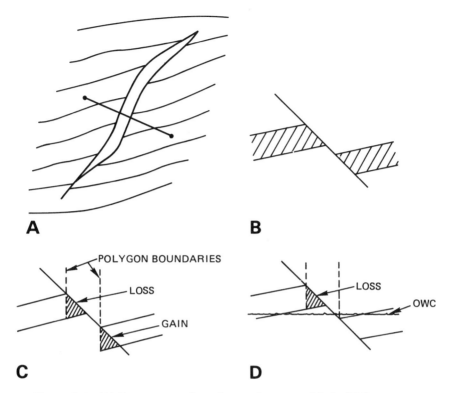

Figure 8.4 *(A)* Contour map of a surface cut by a normal fault. *(B)* Cross section showing both the mapped surface and the surface below. *(C)* Using the same fault-block polygons for both surfaces results in a shift of volumes for the unit bounded by those surfaces, with no significant net gains or losses. *(D)* When the same fault-block polygons are used for both surfaces, and a portion of the unit has dropped below the oil-water contact, a loss in the volume of fluid above the contact will result.

umes result. The shaded portions of the cross section shown in Figure 8.4C represent the volumes that are lost and gained due to the assumption of vertical faulting. It appears that as much volume is gained as is lost, the net effect being a shift of the volume from the gap between the polygons into the down-thrown polygon. No significant net gain or loss in volume results.

However, if fluid contacts, mining benches, or lease boundaries are considered, the effects are less satisfactory. Figure 8.4D shows the cross section of Figure 8.4C, with an added oil-water contact. The volume now desired is that for the oil portion of the rock unit (see chap. 10). In this case, due to the position of the fluid contact, volumes are lost, but never gained. A similar effect is seen when a lease (vertical) boundary passes along the fault gap. In this case volumes would be shifted from one side of the boundary to the other and would not represent the true equity.

149

Different approaches to grid construction may be taken with the fault-block method. The problem of volume distribution differs according to the approach used, and the same evaluation process can be applied to the other methods. Cross sections similar to those of Figure 8.4 demonstrate where volumes are lost or gained in each of these methods.

Method 2. Structural grids are constructed as in Method 1; however, all fault-block grids are combined into one grid. All Null values in fault gaps are replaced by values interpolated from nearby nodes. The result is a single, continuous faulted grid for each surface. Volumes are calculated only inside the fault-block polygons. Method 2 always gives more volume than is actually present (Fig. 8.5A).

Method 3. Structural grids are constructed as in Method 1; however, a new set of fault polygons is used for each structural surface, allowing dip on the fault to be incorporated.

When Method 3 is used, a loss in volume will result if only grid nodes inside the polygons are used to calculate the isochore grid, because the gaps on the upper and lower structural surfaces are not vertically above each other but are shifted so that they are above or below portions of the other surface (Fig. 8.5B). Using only nodes inside polygons means that nodes in the gaps are treated as if they were assigned a Null value. Thus, an isochore built from these grids will contain Nulls where a gap exists on either of the surfaces. Volumes will be lost from both sides of the fault.

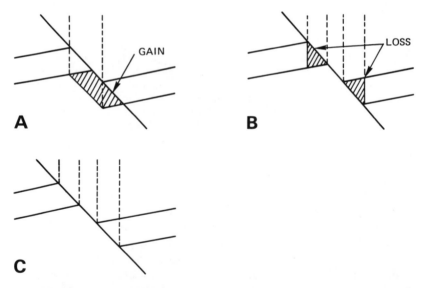

Figure 8.5 (A) Using the same polygon for both surfaces and combining all fault-block grids into a single grid for each surface results in a gain in volumes. (B) Using separate polygons for each surface and calculating volumes only where the surfaces are common to both polygons results in a loss of volumes. (C) Using separate polygons for each surface and combining all fault-block grids for each surface into one grid should result in no gain, loss, or redistribution of volumes.

If grid node values representing the projection of the structural surface past the fault-block polygon into the gaps are used to generate the isochore grids, then isochore thicknesses will project into the gap from both sides of the fault. The decision now must be made as to which set of fault-block polygons — those for the upper surface or those for the lower — to use for volume calculations. In either case, volumes will shift from one side of the fault to the other, similarly to method 1, with no significant net gain or loss in volume.

Method 4. Structural grids are constructed as in Method 1; however, a new set of fault polygons is used for each structural surface. All fault-block grids for each surface are combined into one grid. Values are calculated to replace Nulls in the fault gaps. The result is a single faulted grid for each surface, but now the position of the fault shifts with depth.

When Method 4 is used, there should be no gain, loss, or redistribution of volumes, and the cross section (Fig. 8.5C) indeed seems to show this. However, the process of replacing Null values with calculated values in the fault gaps only approximates the fault plane; the fault-plane part of each grid will be different, so small losses and gains will be distributed along the fault plane.

Methods 1 and 2 require significantly less effort than Methods 3 and 4. Method 4 is the most accurate, with Methods 1, 3, and 2 decreasing in accuracy in that order.

FAULT-TRACE METHOD

Of the four faulting methods in this chapter, the fault-trace method is by far the easiest to use if the required computer programs are available. The other methods combine gridding, grid operations, data calculations, and other mapping functions to produce a final faulted surface. In contrast, the fault-trace technique is a gridding algorithm designed specifically to create faulted surfaces (e.g., Bolondi et al., 1976; Marechal, 1984; Nordstrom, 1985). Data for a faulted surface are converted, in one step, to a continuous faulted grid.

The fault-trace method works with surface data as regular gridding algorithms do. In addition, this method requires information about the fault location to determine which data points are used for calculating grid nodes. In general, the procedure does not allow points on opposite sides of a fault to be used when a node value is calculated. The important aspect of the fault-trace method is that complexities are handled primarily by the program, making implementation simple for the mapmaker.

Defining the Fault Traces

A fault trace represents the intersection of the fault and the surface. If the fault is truly vertical, the intersection will be a single line. The trace in a nonvertical, normal fault is represented on the map as two lines. One of the lines marks the intersection of the upthrown portion of the surface with the fault, and the other line marks the intersection of the downthrown surface and the fault. The surface is not present in the gap between these lines; however, the fault plane is present in this area and connects the separated portions of the surface along the entire length of the fault.

Reverse faults shift one portion of the surface up and over the other, causing two occurrences of the surface at each location near the fault. In these situations, the trace is represented by two lines, one for the upthrown and one for the downdropped fault block. No gap exists, however, and special techniques must be used to map these faults.

Most fault-trace methods require that the trace be defined by a series of points from one end of the fault to the other. If surface data are available only for the area being mapped, then the fault traces need not be extended beyond the map limits. However, if data outside the map area will be used to build the grid, the trace must be extended sufficiently far to ensure separation of data on opposite sides of the fault. For faults that fade out within the map area, the first or last point collected for that trace will be the end of the fault.

A method must be used to distinguish one fault trace from another. Keeping each trace in a separate file is a possible approach, but it is cumbersome when many faults are involved. When all trace data for a surface are in a single file, attach an ID to all points along the traces to separate one trace from another.

Building the Faulted Grid

Since the fault-trace method is actually a gridding algorithm, little effort is required of the mapmaker. Standard information, required when making any grid, plus a few additional instructions unique to the faulting algorithm must be supplied. One of the additional instructions specifies the area of influence of the fault.

A surface may have been so modified by the fault that the opposite sides no longer have any shape relationship. These faults are referred to as major faults. Other faults, especially those having small displacements, may have affected the surface only a short distance from the trace, in which case points sufficiently far from the trace could be used when calculating grid node values across the fault. These faults are referred to as minor faults.

The process of calculating grid node values on a faulted surface is similar to the method used for unfaulted surfaces. The points to be used are selected and weighted in some manner (usually related to distance), and a value for the node is calculated. A major difference in the faulting algorithm is the point-selection process. If the fault is major, the fault trace acts as barrier beyond which the algorithm cannot look for points; this restriction is often referred to as a "line-of-sight" test. A data point separated from the grid node being calculated by a major trace cannot be used.

Figure 8.6A shows a grid node for which a value is to be calculated. Two major

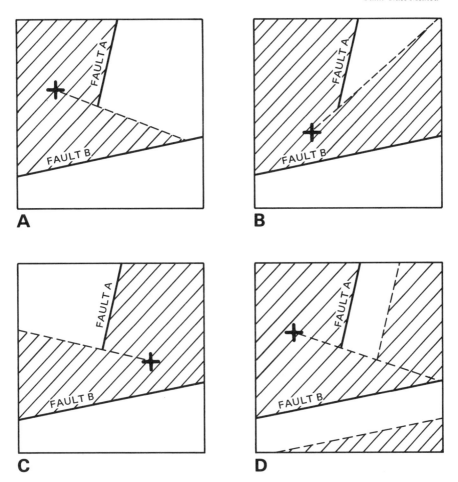

Figure 8.6 Fault A fades out within the map area, and fault B extends across the area. (A, B, C) Faults A and B are treated as major faults and prevent the gridding algorithm from using data on the other side of faults. Only data points within "line-of-sight" of the grid node (+) are used (hatchured area). (D) Faults A and B are treated as minor faults. Data greater than a specified distance from the fault are used to calculate node values on the other side of the fault. Data less than that distance are not used or are discounted.

faults cut through the map area. These faults restrict to the shaded area the data available for calculating that node's value. Grid nodes located to the south and west will be able to use more data from the other side of fault A since the node can "see" more of the area (Fig. 8.6B). The same is true for nodes on the right-hand side of fault A (Fig. 8.6C). Grid-node values will gradually change around the end of a fault, making one continuous surface from one side of the fault around the fade-out to the other side.

153

A minor fault is assumed to have local effect on the surface — that is, the surface is distorted only in the immediate vicinity of the fault — but not distant influence. The general shape of the surface is similar on both sides of the minor fault away from the trace; in contrast, the surface may have a totally different form across a major fault. In this case some fault-trace algorithms allow specification of the width of a zone of influence along the fault trace. Points within this zone are not used for calculating node values on the opposite side of the fault; if they are, a special weighting function is applied to reduce their influence.

Figure 8.6D shows two minor faults and a grid node to be calculated. Data points within the shaded area would be used to calculate the node's value as if the faults did not exist. Data points within the zone near the fault may not be used at all or may be discounted on the basis of a weighting function.

The minor-fault concept should be used with caution, as some undesired results may occur. If few data exist within the zone of fault influence, points on the opposite side of the fault will have significant influence on the calculated surface. This tendency to shift towards the elevation across the fault gives a "rollover" effect, which is generally not desirable. To avoid the problem, use a zone of influence only if there is enough data within the zone on both sides of the fault to define the surface near the trace.

Fault-trace gridding algorithms are usually more complex than those for unfaulted surfaces. These additional complexities are required to calculate node values near fault traces and in V-shaped junctions where two traces intersect. Since the data in these areas are located on only one side of the grid node, the algorithm must extrapolate to assign a node value. In extrapolation, the nearest data points have a significant impact and may cause unstable results. Some algorithms are modified to reduce extrapolations in these areas. Others do not calculate values for nodes in tight V-shaped junctions between fault traces because so few points are available. Assigning Null values to these nodes creates problems for contouring and volume calculations; the surface exists there and should be used in the calculations, but Nulls imply that it does not exist.

The considerations for handling multiple surfaces with the fault-block method also apply to the fault-trace method. If the faults are vertical, the same traces are used for all surfaces. If no growth is associated with these faults, isochore grids are added or subtracted to construct other conformable surfaces in the sequence. If growth is present, each structural surface must be built separately, or the isochore grids must be built using the fault-trace method.

The isochore methods described in chapter 6 for building conformable surfaces cannot be used as described if the fault planes are not vertical because surfaces on one or both sides of the fault will project incorrectly into the gap. Three steps are used to build an acceptable surface: (1) combine the traces from both surfaces to form a "composite" fault gap; (2) set to Null all nodes in that gap; and (3) use the remaining nodes as additional structural data to grid the conformable surface up to

154

the trace for that surface. As this may require capabilities not typically found in mapping programs, an alternative but less desirable approach is to build each structural surface independently.

Applying the Fault-Trace Method

As described above, the fault-trace method uses the fault trace to control grid construction. As a result, the grids contain sharp discontinuities where the faults intersect the surface. However, if these faulted grids are contoured using standard algorithms, the faults will appear as steep-sided folds with rounded edges and a ragged appearance as they zig-zag through the grid nodes. Fault gaps will appear as blank areas where contours stop abruptly at Nulls, and the fault traces will not automatically be drawn on the map.

To avoid these problems, the contouring program must be specially designed to use the fault-trace data. In fact, the fault-trace method is not merely a gridding algorithm but a group of algorithms, one for grids and others for contours, cross sections, volumetrics, and inverse interpolation. All of these tasks must be modified to work with fault traces if this method is to be used throughout a project. Fortunately, the major part of the algorithm is the same for most of these tasks.

Contour Maps

Only grid information on one side of the fault should be used to draw contours in that part of the surface. The same concept of line-of-sight applies to contouring as to gridding. Problems of extrapolating contours beyond grid nodes to fault traces must also be handled. When solutions to these and other problems unique to the fault-trace method have been built into the contouring algorithm (cf. Pouzet, 1980), a good set of faulted contours can be built (Fig. 8.7). These contours will extend smoothly up to the fault trace and around the ends of faults.

Contours of the faulted surface alone are satisfactory for many projects; however, we may wish to contour the fault plane as well. Recall that this is done in the fault-block method by using the gridding algorithm to assign values to grid nodes containing Nulls in the fault gap. With that method the contours approximate the fault plane, but do not portray sharp boundaries and tend to zig-zag through the grid nodes.

Some fault-trace contouring algorithms contour the fault plane as well as the surface. Most do not grid the gap where the fault plane exists, but instead fit a mathematical function to points along the fault trace. The points can be taken from along the intersection of the trace and faulted surface to provide a set of data along both sides of the fault gap. The resulting contours portray the slope and character of the fault plane.

155

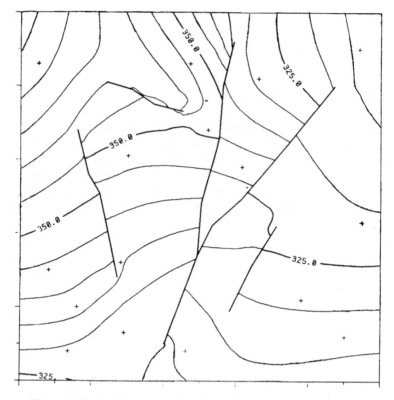

Figure 8.7   Contour map constructed using the fault-trace method.

Cross Sections

Data on fault-trace location should also be used in generating cross sections. Profiles of the surface should extend smoothly up to the trace, and the trace on one side of the gap should connect with the trace on the other side. A straight-line connection may seem adequate for the profile of a fault plane and for many cases will be acceptable. However, for situations where the line of section is nearly parallel to the trace, the fault plane will contain many irregularities and will not plot as a straight line. Therefore, a technique similar to that used to contour the fault plane should be used to generate profiles.

Volumes

Volume calculations with the fault-trace method have the same potential for problems as with the fault-block method. Recall that four situations are possible, depending upon the procedure used (Figs. 8.4 and 8.5). All of the gains, losses, and redistribution of volumes encountered with the four cases of the fault-block

156

method are also encountered with the fault-trace method. The fault-block method requires a large amount of effort to create grids used for the most accurate calculations (Fig. 8.5C). The fault-trace method, however, requires only the specification of a new set of fault traces. The increased accuracy in volumes is usually worth the extra effort.

Fault-trace procedures for calculating volumes vary widely in how they calculate volumes in the area of the fault gap. Some use two mathematical functions to approximate fault planes for the upper and lower surfaces. Theoretically, these fault planes are the same surface and should be coincident if they overlap. This coincidence of surfaces is difficult to achieve because two separate sets of data, one for the upper surface and one for the lower, are used to approximate this surface independently. A possible solution is to use data from both surfaces to make one fault plane surface. However, this becomes extremely complicated when several faults intersect. Even so, potential problems still exist because inaccurately located traces could contort the generated fault plane.

Inverse Interpolation

Inverse interpolation is used in a number of procedures in this book and some of those procedures could use grids built by the fault-trace method. In most cases, the standard inverse-interpolation method used for unfaulted grids is appropriate. However, grid nodes on the opposite side of a fault should not be used to estimate a value at an X-Y location near a fault. Again, an algorithm similar to that used for contouring faulted grids should be used to estimate the surface near the trace.

Reverse Faults

To this point, discussion of the fault-trace method has assumed either vertical or normal faults, for two reasons: First, the discussion is greatly simplified, and second, state-of-the-art fault-trace gridding algorithms are just beginning to address the reverse-fault problem. As will be seen when reverse faults are discussed in the fault-plane method, complexities encountered in gridding reverse-faulted surfaces increase by an order of magnitude. Although the fault-plane method will handle reverse faults, the process is too cumbersome to use with more than a few faults. The advent of fault-trace methods for reverse faults will be a significant and welcomed addition to computer-mapping capabilities.

RESTORED-SURFACE METHOD

The restored-surface method differs from the others described in this chapter in that it allows data on opposite sides of the fault to be used when constructing a grid of a surface. It does this by estimating and gridding the vertical displacement due to faulting. This displacement grid is used to remove the effects of faulting from the

157

data (Sebring, 1958; Powers, 1984). Gridding the adjusted data gives the restored surface – a grid that is meant to estimate the prefaulted surface. The displacement is added back to the restored-surface grid to arrive at the faulted surface.

The restored-surface method requires more definitive knowledge about the fault than do other methods, and so it is not convenient to use with highly faulted surfaces. It is, however, extremely useful as a tool for testing interpretations of correlation, timing, amount of movement, and growth of faults (Johnson, 1977). The prefault surface that is constructed can be checked to see if all throw due to faulting has been removed. If not, throw can be added or deleted, and the surface can be checked again. In projects involving different periods of faulting, the displacement associated with one period can be removed. Isolating one period of faulting from another simplifies the reconstruction of the surface's configurations through geologic time.

Data Collection

The restored-surface method requires the standard (X,Y,Z) surface data used by all gridding algorithms discussed in this book. In addition, it requires information about the location of the faults and vertical displacement at points along the fault traces. Unlike the fault-block and fault-trace methods, which allow the use of two trace lines to define the fault gap, the restored-surface method allows only a single line to be used for the fault trace. Normally the single-line trace is drawn along the center of the gap.

The displacement of the surface at the fault trace is the vertical difference between adjacent points on opposite sides of the fault. This displacement is usually assumed to vary gradationally along the trace, with no sharp discontinuities. Vertical displacement is often estimated from hand-drawn maps. If a well cuts the fault or a seismic line crosses the fault, a more accurate estimate of displacement can be obtained. Obviously, if more points are collected along the trace, more detail can be built into the fault.

This method assumes that all vertical displacement occurs on one side of the fault; we refer to this side as the displaced side. If displacement is upward, the displacement measurements are positive. If displacement is downward, the measurements are negative. This terminology does not say that all displacement is found only on one side of the fault, but the approach is convenient for mapping.

In addition to displacement at the fault trace, the restored-surface method also requires data about displacement of the surface away from the trace. The minimum information required by the procedure is the location of the zero-displacement line (chap. 2). All points in the area between this line and the fault trace have some amount of vertical displacement; all points outside the area have zero displacement. Using the zero-displacement line and values on the fault trace, a grid of displacement can be built. However, displacement data points elsewhere on the active block will further define the displacement grid.

158

In many cases the geologist has little or no knowledge about the location of the zero-displacement line. The authors have a useful but pragmatic method for defining its position in these cases. The method uses displacement at three points on the fault trace — one at each end and one in the middle — to define the position and displacement of a zero-line control point. A reasonable position for the zero-displacement line can be extrapolated using the zero-line control point and the three points on the trace. This point also aids in drawing contours of displacement for the active block.

Figure 8.8A demonstrates the method for a fault that fades out within the map area. Here three points and their displacement have been defined on the fault trace. The fault fades out on both ends, so the displacement at those points is zero. The displacement at the midpoint is −200; it is negative because the block dropped. The zero-line control point is located on the displaced block, along a line normal to the fault trace that intersects the trace midway between the two endpoints. The control point is placed on this line at a distance away from the trace equal to half the distance between the trace endpoints. The displacement at the control point is the average of the displacements at the trace endpoints. In Figure 8.8A, the zero-line control point lies on the zero-displacement line.

Only one end of the fault in Figure 8.8B fades out within the map area. The other endpoint, marked by the intersection of the map and the fault, has a displacement of 200; it is positive because the block moved up. The same procedure is used to locate the zero-line control point and assign a displacement value, in this case, 100. The displacement at a third point along the trace has been defined as 300. Using the three points and the control point, contours of displacement can be

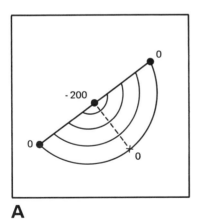

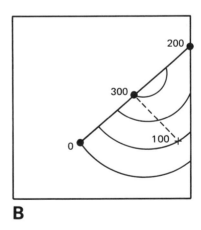

A　　　　　　　　　　　　　**B**

Figure 8.8 *(A)* Displacement contours for a fault that fades out within the map area. The trace endpoints are used to define the zero-line control point. The four points and displacement values are used to draw the contours. *(B)* Displacement contours for a fault that extends outside the map area. In this case the zero-line control point does not lie on the zero-displacement line.

159

approximated and the zero-displacement line defined. The authors have found that a minimum of three points along the fault trace is required to define the zero-displacement line effectively.

Building the Grids

In Figure 8.9A, a set of data were gridded without accounting for the fault. The fault is not sharply defined, but shows up as a fold. The same data were gridded using the restored-surface method to incorporate the fault (Fig. 8.9D). Figures 8.9B and 8.9C show the steps involved in producing the faulted grid. The following discussion outlines the steps required to incorporate the fault:

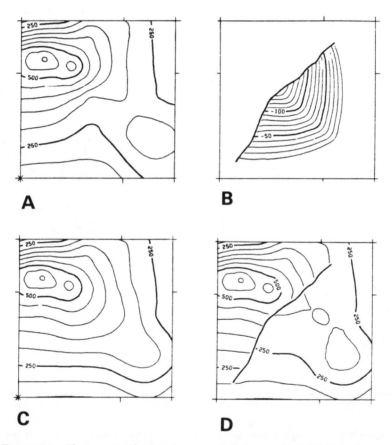

Figure 8.9 The steps involved in the restored-surface method. (A) If the data are gridded without faults, a smooth surface with a fold results. (B) Contour map of the fault-displacement grid. (C) Contours of the restored surface. (D) Contours of the faulted surface.

1. Build a grid of the vertical displacement of the surface due to faulting (Fig. 8.9*B*);
2. Use interpolated values of displacement to restore the horizon picks to their prefault (restored) elevations;
3. Build a prefault grid of the surface using the restored horizon data (Fig. 8.9*C*);
4. Add the vertical displacement grid to the prefault grid to create the faulted-surface grids (Fig. 8.9*D*).

The displacement grid can be built in a number of ways. Perhaps the most obvious is to digitize a hand-drawn map of displacement and grid those points. Another is to locate a number of control points on both the displaced and the nondisplaced portions of the surface and build the grid from these. More desirable would be an automatic technique for generating the displacement grid from any available displacement data, even if only a few points. This technique might generate additional control points that would then be used to generate the grid.

An automatic technique that uses the zero-line control point discussed above might generate a set of supplemental control points (cf. chap. 5), which might take the form of ring points or a coarse grid of points. If two additional points are available, or if the three fault-trace points and the zero-line control point can be used to calculate two points, then a total of six or more points is available. These points can be used to generate a quadratic polynomial (second-order trend surface; chap. 11). Using this polynomial, calculate additional control points, add them to the displacement data supplied by the geologist, and build the displacement grid. Any automatically generated control points that are close to displacement data should be removed.

Building the displacement grid is more complex than it first appears because grid-node values on one side of the fault will have large positive or negative displacements and those on the other side will be zero. A fault-trace gridding algorithm can be used to incorporate this discontinuity but may have problems defining the zero-displacement line if few data are available.

A possible approach is to allow the displacement grid to extrapolate across the fault into the nondisplaced block. If the displacement is downward, the grid will extrapolate negative values, but should be constrained to have a maximum value of zero (assign zeros to all nodes with positive values). If displacement is upward, the grid should be constrained to have a minimum value of zero. This grid now contains zeros in all nodes outside the zero-displacement line and contains positive displacement values on the displaced block and where it extrapolated into the nondisplaced block.

Now we must cut off this nondisplaced portion. One method defines a polygon for the nondisplaced block. Use the fault-trace points as vertices; if the fault fades within the map area, straight-line projection to the map edge defines other vertices. To close the polygon, use the corners of the map on the nondisplaced side of the fault. Compare this polygon to the displacement grid just constructed, and

161

assign a zero value to all nodes within the polygon. Another method that does not involve polygons uses a steeply dipping grid of the fault plane to cut off the nondisplaced portion of the displacement grid.

As the second step, the vertical-displacement grid is used to restore the horizon tops to their prefault position. This is done by interpolating a value from the displacement grid at each well location. These interpolated values are then subtracted from the original picks, removing the displacement caused by faulting. Values interpolated from the displacement grid for wells not located on the displaced block will be zero, so only those picks affected by the fault will be altered.

Picks near the fault may create problems for the inverse-interpolation algorithm. Grid nodes will have values of zero on the nondisplaced side and values that are positive or negative on the displaced side. If nodes from both sides of the fault are used for interpolation, an intermediate value will be calculated. This value will be incorrect, and the horizon pick will not be totally restored (if on the displaced block) or will be shifted when it should not be (if on the nondisplaced block). There are at least two ways to prevent these problems. A procedure for inverse interpolation similar to that used for the fault-trace method could be used. That method uses nodes from only one side of the trace to interpolate the value. Another approach uses the displacement grid that extrapolated past the fault. Only those horizon picks on the displaced block (outside the polygons used to modify the displacement grid) should have values interpolated.

The third step is to build the restored surface grids. The restored horizon picks, if the displacement grid is correct, should be at their prefault elevations. Therefore, points on opposite sides of the fault can now be used to construct the restored surface grids. Thus, even fault blocks with very little (or no) data can be gridded. The restored surface should be mapped and carefully examined for geological reality. If the generated grids are complex in areas near the faults, it is possible that too much or too little displacement has been removed. In that case, the displacements should be modified and these three steps repeated.

In the final step, the fault displacement is put back into the surface by adding the displacement grid to the restored-surface grid.

Handling Multiple Faults

Normally more than one fault must be built into a surface grid. The general procedure is to build a displacement grid for each fault and to combine the grids. The combined grid is used to restore the original data and then to displace the restored-surface grids. Often, however, this process is not a simple addition of displacement grids.

When multiple faults occur, they may be either independent or dependent. Two faults are independent if a fault's effect on a surface is not altered by the presence of another fault. For example, Figure 8.10A shows two crossing faults. The displacement associated with each is indicated at points along the traces of the

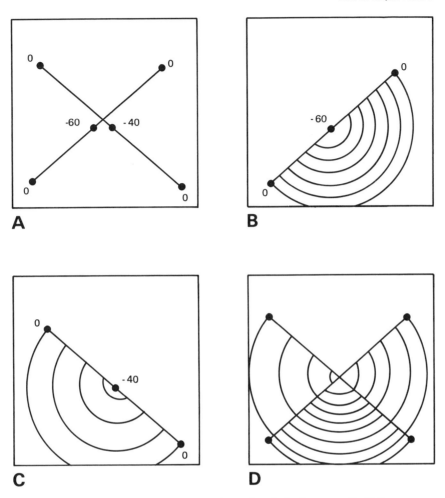

Figure 8.10 *(A)* Fault traces for two independent faults that cross. *(B, C)* Contours of displacement for each fault are constructed independently. *(D)* The displacement contours are added to produce the combined displacement contour map.

faults. The faults are independent, so their displacement grids are built and summed to produce the combined effect of both faults (Fig. 8.10*B*, *C*, and *D*). Dickinson (1954) discusses the problems associated with intersecting faults and their effect on surfaces.

Two faults are dependent if a fault's effect on a surface is affected by another fault. Figures 8.11*A* and 8.11*B* show a map and cross section through a graben. The method for defining the fault-displacement grid associated with fault A would affect portions of the surface beyond fault B (Fig. 8.11*C*). Similarly, fault B affects portions of the surface on the other side of fault A. The area of displacement of one

163

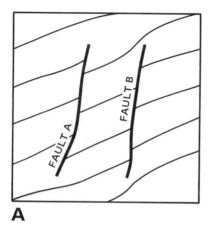

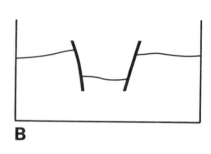

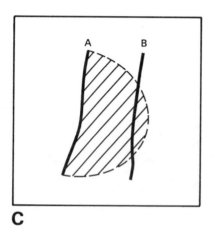

Figure 8.11 *(A)* Contours of a surface cut by faults A and B. *(B)* Cross section cutting both faults. *(C)* The displacement of the surface due to fault A is dependent on the position of fault B, so the displacement due to fault A should not extend past fault B.

fault depends upon the position of the other and should be restricted to the area between the two faults.

Dependent faults often create problems when they are combined to form the total fault-displacement grid. If the displacement grid for fault A (Fig. 8.11C) is operated against a polygon defining the nondisplaced block of fault B, that portion of the displacement grid can be set to zero values. A similar process can be performed on the fault B displacement grid. Both displacement grids have now been modified to account for the other fault.

However, if the displacement grids are added, the displacement along the trace of fault A where the displaced block of B contacts it will be increased to more than was originally observed. A similar situation occurs for displacement along the trace of fault B. It is usually easier to build a single grid that accounts for the displacement on all faults that are dependent. The same procedures used to create

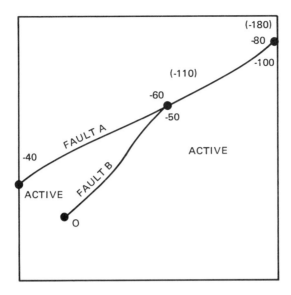

Figure 8.12 Intersecting faults that do not cross can be treated as independent. Where the two faults have merged, the displacement caused by each fault must be specified.

the single-fault displacement grid described above can be modified to work in this situation.

Two faults that intersect but do not cross (Fig. 8.12) could be treated as either independent or dependent faults. The simplest approach is to treat them as independent. In this example, faults A and B intersect, with fault B cutting the displaced block of fault A. The displacement at the fault trace to the left of the intersection is significantly less than the displacement to the right of the intersection. To the right, a portion of the displacement belongs to fault A and a portion to fault B. Fault trace A is defined with points having displacements of -40, -60, and -80, and trace B is defined with points having displacements of 0, -50, and -100. Fault-displacement grids are built separately for each fault and then added to produce the total displacement grid.

Growth Faults

Growth faults, as described in chapter 2, are faults that continue to move through a period of geologic time — that is, deeper surfaces exhibit more displacement than shallower surfaces. Because the restored-surface method builds a grid of displacement, that displacement can be distributed through time, allowing growth faults to be modeled. In this case, the lowest horizon has experienced the most

165

displacement (by definition 100%), and that amount is used to build the displacement grid for the fault.

A portion of the total growth on the fault will take place during the time between the formation of the lowest surface and the next-higher surface. If, for example, 20% of the total growth occurred during that time interval, the displacement grid would be reduced (each grid node multiplied by 0.20 – that is, 20% of the total displacement) and the resulting grid used to restore horizon picks for that surface. The process is repeated upwards until the top surface is reached or until the cumulative growth reaches 100%. If 100% is attained – that is, the displacement grid is depleted – surfaces above this horizon were not affected by the fault, as fault movement ceased at that geologic time.

Timing of Faults

The timing of the fault in the geologic record must be considered when using the restored-surface method to build grids of the stratigraphic framework. Figure 8.13 shows a reverse fault cutting surfaces A and B and terminating at surface C. The fault occurred after the creation of surface B and before the creation of surface C.

Deciding when a fault occurred or what surfaces it affects is normally easy. Understanding how to incorporate the fault into the grid-building processs is not always as easy. To help in this understanding, the framework table of chapter 6 can

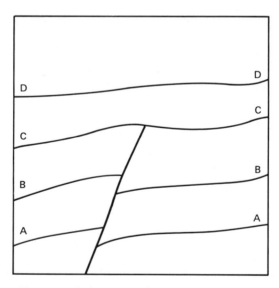

Figure 8.13   The reverse fault cutting surfaces A and B and terminating at surface C occurred after the creation of surface B and before the creation of surface C.

be expanded to include faults. This requires adding another column between the "sequence control" and the "initial grid-building method" columns. Label this column "fault period." Numbers will be entered in this column to indicate the period of faulting. The number of the fault period should be entered on the line of the youngest (shallowest) surface affected by that period of faulting. Another list or table should be made containing all the faults that occurred during each period of faulting and the fault sets to which they belong.

The fault-period column is positioned in the framework table to indicate that data for each faulted surface must be restored before the initial grid of that surface is created. After initial grids have been constructed, the fault-displacement grids are added to all surfaces affected by the fault. After the displacement has been added, the truncation and baselap operations are performed.

For faults with growth, additional columns must be added to the table. One column is used for each fault (fault-displacement grid if several dependent faults are built into one grid) or group of faults that experienced the same proportion of growth. The percent of the total displacement experienced by each surface is recorded.

Applying the Restored-Surface Method

With this method, a single grid is built for each surface. No fault gaps were built into the grids, so no Nulls are found in those areas. Normally, these grids are used directly for contour maps, cross sections, volumes, and inverse interpolation. Contour maps generated from these grids will not have sharp boundaries separating one fault block from another. Instead, faults will be represented as steep slopes, the width of the sloping fault being approximately one grid interval. This is because nodes on one side of the fault are offset in elevation from those on the other side. After the grid has been constructed, nodes along the fault trace can be set to Null values. Doing so will prevent contours from extending across the faults (Fig. 8.9D) and will produce a more pleasing map. A simpler approach uses the fault-trace contouring procedure to prevent contours from crossing the faults.

Cross sections are easy to generate since only one grid is displayed for each surface. These sections, as with contour maps, show a steep slope over the fault cut as the line of profile jumps from one side of the fault to the other.

Because vertical faults are assumed, volumes are redistributed in a way similar to that described for the fault-block method and vertical faults. Also, since the gap was not built into these surfaces, there will always be a gain in volumes for faults that should have had gaps. Since the restored-surface method as described here works only with vertical faults, the discussion related to volume calculations for shifting traces does not apply.

If algorithms for the fault-trace method are available, they can be used for display and calculation with grids from the restored-surface method. The same trace locations used to generate the grids are used to control the fault-trace

167

algorithms. The displays will have sharp boundaries at the faults, and interpolation will be accurate near the fault; however, volumes will still be shifted and show a gain in the areas of fault gaps.

FAULT-PLANE METHOD

The fault-plane method involves building grids of each fault plane and of each surface on both sides of every fault (Peikert, 1970). The fault-plane grids are then used to restrict the surface grids, which will have extrapolated past the fault, to the appropriate side of the fault. The method is designed specifically for dipping fault planes and works with either normal or reverse faults.

Although this method is potentially able to incorporate an unlimited number of faults, it is not practical for more than a few faults. It has several other characteristics that restrict its use: (1) enough data must exist to grid the fault planes properly (Reiter, 1947); (2) special processing is required to handle faults that fade out within the map area; and (3) data on each fault block must be sufficient to build a grid for that portion of the surface.

Figure 8.14A shows in cross section a surface cut by a reverse fault. The fault has

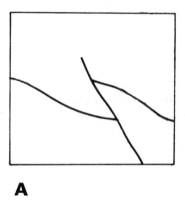

A

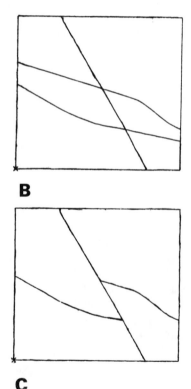

B

C

Figure 8.14 Cross section showing reverse fault. *(A)* A surface cut by a reverse fault is shifted up and over itself. *(B)* Separate grids are built for each side of the fault, using only data from the appropriate side. These grids will protect past the fault. *(C)* Grid-to-grid operations between the fault grid and the surface grid prevent the incorrect projections.

168

shifted the surface up and over itself so that it is repeated vertically in the central portion of the figure. A single grid cannot hold all the information about the repeated parts of the surface. However, two grids can be used to define the surface, one for each side of the fault. Applying the method to this example involves the following steps:

1. Build a grid of the fault plane;
2. Build a grid for the portion of the surface to the right of the fault; note that this initial grid will incorrectly project past the fault plane (Fig. 8.14B);
3. Build a grid for the portion of the surface to the left of the fault (this grid also projects past the fault plane);
4. Use grid-to-grid operations for truncation and baselap to prevent the two surface grids from projecting past the fault-plane grid (Fig. 8.14C).

Data Collection

The fault-plane method requires data on both the surface and the fault plane that cuts it. As with mapping frameworks, a pick on the stratigraphic surface is distinguished from a pick on the rock top in the fault plane (Iglehart, 1970), and these two types of information must be gridded separately. Since this method actually builds a fault plane grid, the (X, Y, Z) locations of intersections of the well bores with the fault plane, or other information about the three-dimensional location of the fault plane, are required.

Since a different grid is constructed for the surface on each side of the fault, picks from each side must be stored in separate fields in the data files corresponding to each block. In general, for a given surface, the number of fields will equal the number of portions of the surface isolated by faults. This data file configuration can be used for both normal and reverse faults. However, normal faults do not repeat a surface vertically, so data are often recorded as one field per surface, but eventually must be separated into one field per fault block. For large amounts of data, manual separation into two fields can be tedious. The following steps can be used to automate this procedure.

1. Build a grid of the fault plane.
2. At each X-Y location in the data file, interpolate an elevation from the fault grid. Combine the interpolated value with the data file as a new field.
3. Replace the Null values in the field containing the original fault data with the interpolated values. Do not change any of the original values.
4. Create two new fields for each surface. Place all values that are lower than the value in the fault field in one of the new fields and values that are higher in the other. Set all other values in these two fields to Null.

The data is now in the general form described above. This automatic procedure becomes more complex with more faults.

Building the Fault-Plane and Surface Grids

The fault-plane grid and grids for the surface on each side of the fault are built using the techniques discussed in chapter 4. All of these grids, whether for fault planes or surfaces, should have the same areal coverage and the same distribution of calculated nodes because grid-to-grid operations will be performed. Grid extrapolation can be a problem, especially for surfaces with limited data distribution, as is often the case for fault planes or isolated fault blocks.

The most common extrapolation problem occurs in areas where the dip and strike of the fault approach that of the surface it cuts; either the surfaces or the faults can change their projection slope and recross one another several times. Additional control points may be needed to prevent recrossing (chap. 5). The authors have found an automated procedure that works well: Produce a trend surface grid, interpolate ring points from the trend grid, and combine those points with the surface data.

Once the initial grids are constructed, the next step is to perform grid-to-grid operations to prevent the surface grids from projecting through the fault-plane grid. The operations used in chapter 6 to create baselapping and truncating relationships are used, and the fault plane is treated as if it were an unconformity. As in building a stratigraphic framework, the process starts with the deepest grids and works upwards. The surface grid lower than the fault plane (upthrown for a normal fault, downthrown for reverse) is compared to the fault-plane grid, and a new lower grid, the minimum of the two, is created. The surface grid above the fault plane (downthrown for normal faults, upthrown for reverse) is compared to the fault-plane grid, and a new upper surface grid, the maximum of the two, is created. Instead of projecting past the fault, the surfaces on either side now project to the fault and then follow the fault plane.

The discussion so far has described the fault-plane method as applied to one surface and one fault plane. Processing becomes complex rapidly as more surfaces and faults are added. Consider the case of one fault and many surfaces. The surfaces may all be conformable or may have truncating or baselapping relationships. The presence of the fault has separated the surfaces into two sets. Within each set of surfaces, stratigraphic relationships are still the same, so the framework procedures discussed in chapter 6 are used.

1. Separate the data for each surface into separate fields, one for each side of the fault.
2. Using the data for one side of the fault, follow the complete framework-building process described in chapter 6. Check the surfaces to see that incorrect extrapolation does not cause the grids to recross the fault plane. Repeat for data on the other side of the fault. Some of the higher or lower surfaces may be present on only one side of the fault; the framework procedure can be altered to account for this.
3. Once the framework for surfaces on both sides of the fault has been built,

truncate the lower-surface grids (upthrown for normal faults, downthrown for reverse) with the fault-plane grid.

4. Use the baselap grid-to-grid operation to prevent the upper-surface grids (downthrown for normal faults, upthrown for reverse) from projecting past the fault plane.

When more than one fault is involved and the faults do not displace one another—that is, they may merge but do not cut and displace each other—processing is basically the same as for a single fault. Now there will be one surface field for each fault block. Grids within each fault block are built using the stratigraphic framework procedure of chapter 6. Once all the surface grids are constructed, they progressively undergo baselap and truncation operations by the fault-plane grids that bound them until all surfaces have been constrained to the fault block in which they belong.

A fault that cuts and displaces another fault adds complexity. In this case, data for the fault that has been displaced must be separated into two fields, just as was done for the faulted surfaces. Separate grids are built for the displaced fault, one for each side of the cross-cutting fault. Once all the surface and fault grids have been built, they must be operated against each other.

The cross-cutting fault is operated against all grids on either side of it, including other faults that it cuts. These are the minimum and maximum grid-to-grid operations appropriate for the particular fault/surface relationship. Once operations with cross-cutting faults have been performed, all the other fault/surface operations are done.

This description of the technique is generalized to handle any cross-cutting fault situation. In most cases, simplified relationships between faults allow fewer grid operations than indicated here. Again, the entire process requires that adequate data be available within each fault block and for each fault plane to build the grids properly.

Applying the Fault-Plane Method

Because each surface is defined by many grids (one on each side of each fault), calculating volumes can be complex. The simplest approach is to generate a set of isochore grids for the unit of interest by subtracting grids for the top and base of each portion of the unit that is isolated by faults. The isochores are then summed and the volumes calculated.

The continuous-surface grids have a drawback similar to those built in chapter 6 for the stratigraphic framework. The grids can be used for calculating volumes, but they will not be accurate where surfaces intersect and become coincident (discussed in chap. 9). Methods to correct this problem involve using the grids before they are made coincident—that is, using the pre-operated crossing grids. Methods described in chapter 10 involve defining an envelope of the top and basal limits of the interval and then subtracting these limiting surfaces to obtain a grid to be integrated.

If contour maps of particular surfaces are desired, procedures similar to those described in chapter 7 are used; the grids that cross before operating with the fault-plane grids isolate the portion of the surface to be contoured. That process must be repeated for each grid that represents a part of the surface.

Isochore grids can be generated from the final grids by the same technique used above for calculating volumes; however, these grids will have zeros at all nodes where the unit is not present. When the grid is contoured, the zero contour should not be displayed, as it will tend to wander through the area of zero-valued grid nodes. Instead, a special contour (e.g., 0.01) can be used to indicate the edge of the unit (see chap. 7).

Faults That Fade Out within the Map Area

A fault that fades out within the map area presents a problem because the surface on one side of the fault is attached continuously around the fault to the surface on the opposite side. We use a grid to represent the fault plane, so it must be allowed to extrapolate across the entire map area, even though the fault actually ends. A procedure similar to that used in the fault-block method is used to ensure that the surface on one side of the "zero-throw" portion of the fault plane smoothly connects with the surface on the other.

Figure 8.15A shows contours of a surface cut by a fault that fades out within the map area, as well as the data available for the surface and the fault plane. The following steps would be used to build grids.

1. Define a polygon that outlines the area where the fault does not affect the surfaces (hatchured area in Fig. 8.15B). This is not the same as the zero-displacement line of the restored-surface method, but is the area defined by a straight line normal to the fault trace that passes through the point where the trace faces out. Everything on the side of that line opposite the fault should be inside the polygon.
2. Grid the surface using all available data, regardless of which side of the fault a point is on, but assign values only to grid nodes within the polygon.
3. Convert the grid values to data and combine them with the two fields containing data for the surface on either side of the fault. This means that the same set of data values is placed in both of that surface's fields (more than two if several faults are involved).
4. Build a grid of the fault plane.
5. Interpolate an elevation from the fault grid at each X-Y location in the data file. Add the interpolated value to the data file as a new field.
6. Replace all Null values in the field containing the original fault data with interpolated values. Do not change any of the original values.
7. Compare the field for the surface above the fault to the field containing the

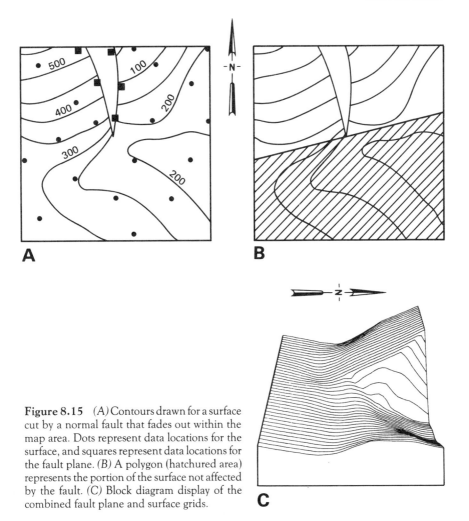

Figure 8.15 *(A)* Contours drawn for a surface cut by a normal fault that fades out within the map area. Dots represent data locations for the surface, and squares represent data locations for the fault plane. *(B)* A polygon (hatchured area) represents the portion of the surface not affected by the fault. *(C)* Block diagram display of the combined fault plane and surface grids.

fault data. Set all surface values that are lower than fault values to Null. Make the same comparison for the field of the lower surface, this time setting all surface values that are higher than the fault values to Null.

8. Use the general fault-plane method as described earlier to build the final grids. Figure 8.15C shows a view of the combined surface and fault planes.

When the final surface grids are created, many of the points used in gridding are from the grid-node locations, and ideally the new grid should duplicate these node values. Therefore, a special program option (Appendix A) that directly assigns a data-point value to a grid node when the two are coincident is valuable and should be used.

173

CHAPTER 9

Forcing
Interpretation
into Grids

The bulk of this book is concerned with placing geological interpretation into grids and maps. Our starting point in chapter 4 introduced the concept of a grid, its creation, and how to use it to generate a contour map without geological considerations. This simple grid contains only such interpretation as that induced by the choice of gridding algorithm or parameters. Such grids are used routinely, but many projects require the incorporation of additional interpretation.

Chapter 5 presents some special characteristics of geologic data and discusses procedures to make appropriate grids. Chapter 8 describes how to create a grid containing interpreted faults. Turning from individual grids, chapters 6 and 7 discuss making grids and maps of several surfaces that are stratigraphically consistent with the interpreted geology.

Rather than being primarily concerned with general features (e.g., data from different sources or the truncation of one surface by another), this chapter is more specific about definite characteristics or interpretations to be incorporated into a single grid. The three sections of this chapter cover making a grid that must be a duplicate of some existing contour map, forcing an isochore zero-line (or similar contour) to have a specified form, and putting a directional trend or bias into a grid.

DIGITIZED CONTOUR MAPS

A common request is that a hand-drawn contour map be "put into the computer." Such a map or request can have a number of possible origins. For instance,

175

a map may have been drawn and agreed upon through negotiation by joint owners of an economic property. It would be unlikely that a set of gridding parameters could exactly duplicate the original map, but because even small variations may have significant economic impact, it is necessary that a grid be created that will represent the map as closely as possible.

Another example is the case of an area that has too few data points for gridding to make a realistic grid. A geologist would contour a map by hand and then introduce it into the computer for further processing. A third example is a map from an outside source, say, a published journal article; here the original data may be unavailable. And finally, we must include the case in which an old grid file was accidently erased and no one remembers the complex series of steps required to generate the final map.

The common aspect of these examples is that an existing map must be converted to grid form, but the grid cannot be generated directly from data. This process of putting a hand-drawn contour map into the computer is referred to here as "digitized mapping," although the term could also be applied to points or lines on a base map. It consists of converting points along the contours to (X,Y,Z) values and then using these values to create a grid. The Z-value is known at all points along a contour, but manual determination of the (X,Y) locations of points on the contour is tedious. For this reason, mechanical devices—digitizers—are widely used to put points along the traced contour line into a computer-usable form.

The creation of a grid through digitizing involves the following steps: preparation of the hand-drawn contour map, mechanical digitization, gridding the digitized data, contouring the grid, and validation by comparison to the original map and digitized points.

Map Digitization and Gridding

It is generally necessary that a map submitted for digitization undergo some preparation for the mechanical digitizing process; seldom will any map be unambiguous or contain all relevant contours. A typical first pass is used to check that all contours are labeled legibly, are clearly drawn, and do not contain such inconsistencies as crossings or abrupt terminations. This type of checking, while obvious, is overlooked in many cases.

A second type of map preparation includes labeling contours. An untrained clerk or technician may be operating the digitizer, so the values of contours must be clear, even if this means marking some contours that were not previously labeled. Labeling is especially needed near closures or in regions where slopes reverse direction. In addition, the contours must be well marked if the contour interval changes on the map.

A third aspect of preparation requires reference locations and coordinates to be specified on the map. The mechanical digitizer, discussed below, operates in terms of arbitrary units (e.g., centimeters or inches), and it is necessary that information

176

be established that allows a given location on the map to be converted to true (X,Y) coordinates. This is done by marking several points on the map, normally near the corners, where the (X,Y) locations are known. These points serve as references for conversion of the digitized points to true coordinates. Two or three points may be adequate for a given digitizer, but more points allow an accuracy check.

Extension of contours beyond the edge of the mapped area is an extremely important aspect of map preparation. Many problems that we see are caused by stopping the digitizing process where the contours end, even if still within the final map area. Gridding these digitized data will usually result in errors at the ends of the contours, particularly in regions where the surface is curved.

Contours should be digitized at least as far as the grid is to extend, and ideally should extend beyond the edge of the map. If done so, the grid will project contours up to the edge without abrupt changes in direction. Extending contours is also important if the contours are within a fault block (chap. 8, fault block method). Extending the line a few grid intervals beyond the faults will improve results. If the method used to grid faults does not allow projection of data beyond the fault trace, additional control along the trace may be needed.

Finally, map preparation includes selection of which contours are to be digitized and which skipped (in an area with dense contours) or determining whether additional contours are needed (in an area with sparse contours). A general rule of thumb is that the contours selected to be digitized should provide constant density over the map. Avoid large empty areas, and digitize only some of the lines in steeply dipping areas. In addition to contour lines, individual points can be added as needed; these typically include well locations and interpreted crests and troughs of folds.

The prepared map is next digitized. Several types of digitizer are available, but most consist of a table on which the map is placed, a stylus or cross-hair with mechanical, electrical, or other linkage, and mechanisms to detect the location of the stylus and convert it to absolute spatial-table coordinates. Reference points with known true coordinates are used to define orientations and scales that convert all digitized absolute locations to map coordinates. These (X,Y) coordinates and associated contour Z-values are then sent to the computer in a form suitable for gridding.

The stylus is normally moved over the contour lines by an operator, although automatic scanning or line-following digitizers are now available (e.g., Dorn, 1983). In either case, the stylus traces a contour line, and (X,Y) coordinates are determined at a number of locations along the line. These points may be selected by the operator, but usually are based on a constant distance increment (e.g., every centimeter along the mapped line). Sometimes more points are selected near curves.

The file created by the digitizing process is then used by the mapping program to create a grid. This can be done by usual gridding algorithms, as summarized in chapter 4. However, additional information is present because groups of points can be connected with lines of constant elevation. Special gridding algorithms can take

177

advantage of this information, although such methods are normally not necessary.

An important consideration here is selection of the grid interval. It should not be larger than the smallest features to be retained on the final map. Small closures or deflections on contours cannot be modeled unless they are larger than the grid interval. However, the interval should not be made so small as to waste computer resources. In most cases, the grid interval is determined by other considerations in the project.

The spacing between contour lines selected for digitizing should not be made too small; some contours should be skipped in regions where they are closely spaced. Very dense data wastes computer and digitizer time, and most gridding algorithms essentially average closely spaced points, thereby flattening slopes. The spacing between points on a line should not be too small, but this spacing is usually made smaller than that between lines. Ideally, there should not be more than one or two digitized points per grid-square.

In short, the procedure to follow is to first review the map and decide on the size of the smallest significant feature. The grid interval should be selected to be about one-half the size of this feature. Then select the spacing of the digitized points along the contour lines to be about one-half the grid interval. This procedure generally gives good results.

Digitized mapping appears to be a straightforward process, but even the experienced practitioner can be surprised by results. The grid should therefore be validated through contouring. This generated map should be the same scale as the original so it can be overlaid and compared directly. Posting the positions of the digitized points can aid in determining why contours stray from their expected locations.

If the two maps do not agree, the geologist must decide if the discrepancies are significant enough to merit additional effort. Assuming that the deviations are not caused by mistakes, the data can be corrected or modified by deleting digitized contours, digitizing additional contours, or adding individual points. Errors can be reduced if certain pitfalls are avoided.

Potential Pitfalls

Certain pitfalls are common when digitizing and gridding contour maps. Typical problems involve selecting the contours to be digitized and determining whether special contours need to be added in empty areas, spacing of digitized points along the contours, and the grid interval to be used. This section presents several examples of maps that point out these difficulties. The digitized points are posted on the maps, with contours from the resulting grids overplotted.

Figure 9.1 shows a broad open area in which the 105-contour wanders somewhat aimlessly from top to bottom. The 110-contour is digitized to the right, and 100- and 103-contours are digitized to the left. However, the blank area is large and

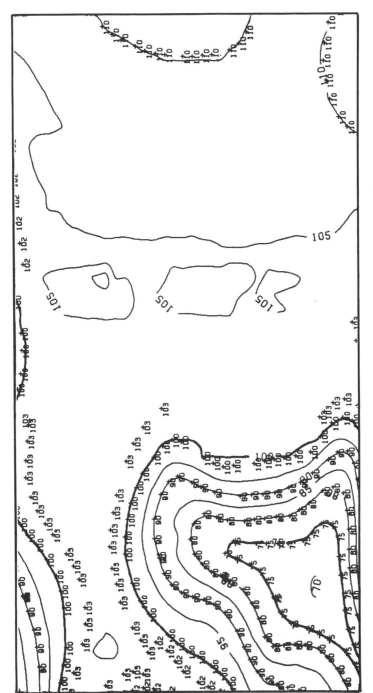

Figure 9.1 Contour map from digitized data, showing the effects of lack of control in flat regions.

flat, so the surface is uncontrolled. Digitizing the 105-contour is necessary, and additional contours should be included.

As a similar matter, note the small 70-closure in the lower left of the figure and the small unlabeled closure (representing 105) in the upper left. Neither of these closures is valid; the original map did not contain them, and they are projections caused by steep slopes of surrounding data. To prevent the 70-closure from appearing, a few points could be placed in that open area with values of 71. Similarly, a short 104-contour or a point would prevent the 105-closure from appearing.

This map points out the fact that most gridding algorithms are sensitive to changes in surface slope and data distribution. Because more contours are drawn in steeply dipping areas, the data density is high and a good fit to the map results. In flat areas, the widely spaced contours lead to low data density and a poorer fit. Accordingly, supplemental curves or points are necessary in flat areas.

Figure 9.2A shows digitized 75-values, but the 75-contour does not follow them into the narrow zone. This is a case where the grid interval is too large to reproduce the narrow feature. Because both 80- and 75-values are found nearby, the node values in that region are essentially local averages, giving such values as 77 in that region. Similar difficulties cause the 90-contour to be erratic, an unexpected 95-contour to be added, and the 100-contour to be inaccurate.

Figure 9.2B shows the same area, but gridded with an interval half that of Figure 9.2A. Here the 75-contour is improved, though at least one node value is incorrect. Results might have been improved if the 80-contour had not been included in the area to the south where contour lines become close together, thereby preventing the 80 and 75 values from being averaged.

Figure 9.3 shows an area in which virtually all possible problems occur, as the gridded contours do not follow the digitized points at all. One difficulty is the grid interval, which is much too coarse for details in the map. Also, study of the resulting 50-contour shows problems with consistency, as it was apparently intended that the 50-contour be included in the central closure. This contradicts the form of the general slope and the digitized 55-contour, although another digitized 50-contour to the right of the 55-contour might have helped. Figure 9.3 appears to be an inadvertant example of attempting to create an incorrect map. No matter how much effort is spent, it will be impossible to create a grid or map that honors incorrectly drawn contours.

Interactive Grid Editing and Mapping

In the past few years, interactive computing has become increasingly common. A typical application displays a contour map on the computer screen on which the geologist analyzes the contours. If the map is not correct, contours are "erased" and new lines sketched in their place with a light pen, thumbwheels and cross-hairs, mouse, or other similar device. The contours are then translated to a form that the

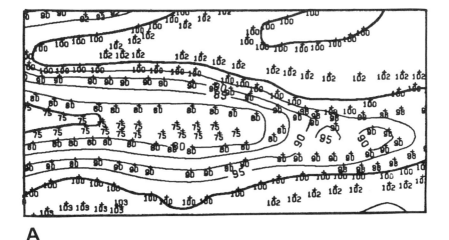

A

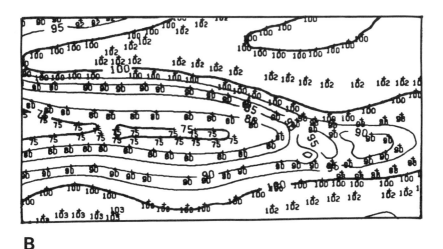

B

Figure 9.2 Contour map from digitized data, showing the effects of digitizing contours that are too closely spaced. *(A)* Grid interval too coarse. *(B)* Grid interval one-half that of *(A)*.

computer can use to generate a new map, either through immediate regridding or through creation of a correction file to be used later.

This process is thus an automated or interactive version of digitized mapping. Indeed, interactive processing is commonly used to validate and correct digitized maps. Slight errors in the grids can be modified directly to have the specified form, saving the time needed to reprocess them with the digitizer.

An obvious advantage of such interactive mapping is speed. The grid or map

181

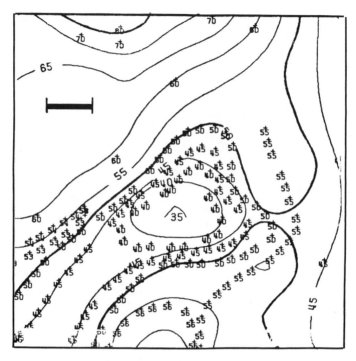

Figure 9.3 Contour map from digitized data, showing the effects of grid interval (indicated by bar) too coarse for desired features, digitized data too closely spaced, and possible inconsistency in drawn contours.

can be evaluated rapidly, and corrections can be quickly entered and incorporated. In addition, the geologist can consider several alternative approaches to the map, drawing and redrawing portions until they seem correct.

A disadvantage of this approach for most interactive programs is that the changes are not saved and are not exactly reproducible. For instance, extensive modifications may have been made to a grid interactively. If another data point is acquired later, the grid would have to be regenerated using the original data and new point, and the modifications entered again. However, there is no assurance that the same grid could be reproduced away from the new data point. Great effort may be required to replicate the map form, and even then the grid values may vary somewhat.

At least two methods may be used to reproduce the map. For the first method, the previously modified grid is converted to (X, Y, Z) data (grid to data, Appendix A), and the new data points are added to this set. After removing those grid-data points near the new points, the grid is generated normally. An alternative method is to use Data-merge (chap. 5). The previously modified grid is converted to data as above and assigned to dataset A, and the new and original data points are made

182

set B. The two datasets are merged, honoring B. Either of these approaches could be done directly by an interactive program.

CONTROLLING THE ZERO-LINE IN ISOCHORE MAPS

Isochore maps are widely used for geologic interpretation, so it is important that they be drawn correctly. Normal methods of gridding will usually be adequate in regions where the unit exists (i.e., has positive thickness), but they can break down at the zero-contour, or zero-line, where the unit pinches out. This section describes steps that can be taken to ensure that the zero-line is reasonable or has a desired form.

Negative Values in Isochore Grids

For most applications, thickness grids should be allowed to project to negative values where the unit being mapped is absent. Since the unit physically has zero thickness in those areas, we might consider that all Z-values in the grid should equal zero in that region. While this is true for a few applications, most uses of the grid will be improved if the grid extends to negative values where the unit is absent.

For contouring, better definition of the zero-contour is obtained if the grid projects through the zero-line to negative values. This continuous projection establishes the zero-contour line from both positive and negative sides. Figure 9.4A shows a profile through a grid of thickness. As a result of the projected negative values establishing the gradient, the zero-contour is located at the arrow. If this grid is contoured with the restriction that the minimum contour level plotted is zero, a reasonable map will result (Fig. 9.4B).

On the other hand, a grid that contains a mass of zero values has less control. Figure 9.5A shows a profile through a grid with many zeros. Here the contour must fall on the grid intersection at the arrow. Many algorithms for plotting contours could not draw the zero-contour because no intersection value is negative, as needed to detect the zero-line crossing in the grid-square. Further, an unjustified change in slope near the zero-line is implied. Simple use of a plotting routine with this grid would not generate a good map (Fig. 9.5B).

Similar considerations apply if the grid is used to integrate areas or volumes, as discussed in chapter 10. The grid in Figure 9.4A will give more accurate results near the zero-line than will the grid in Figure 9.5A, as indicated by the locations of the arrows restricting the integration. Integration above zero will calculate volumes or areas in the positive portion of the grid, but this portion will not be defined correctly. In fact, use of the grid in Figure 9.5A will always give a slightly greater volume, as indicated by the shaded area.

Some applications of grids require that zero values rather than negative values be used (for instance, when adding several isochores to get a total isochore). If so,

183

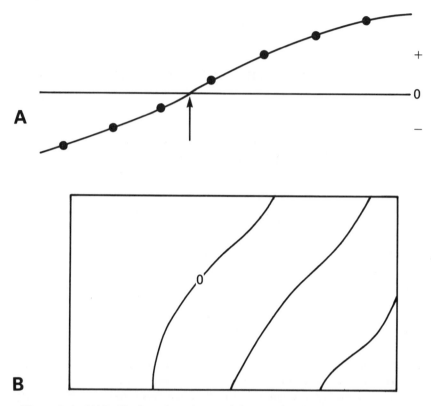

Figure 9.4 *(A)* Profile through isochore grid showing the effect of allowing the grid to take on negative values in the region where the mapped unit is absent. Dots represent grid intersections and values; arrow represents location of zero-line. *(B)* Map corresponding to *(A)*.

new grids are created in which the negative values are changed to zero. Alternatively, restrictions can be included in programs to compensate for negative isochore values. For instance, the routine used to add two grids could treat negative values in the summed grids as zero. This approach can also be used when building conformable horizons.

Data Collection and Gridding Procedures

Even if a grid containing negative values is created, there is no assurance it will give a reasonable zero-contour when mapped. The location or form of the mapped pinch-out could be unreasonable, and manual modifications may be

184

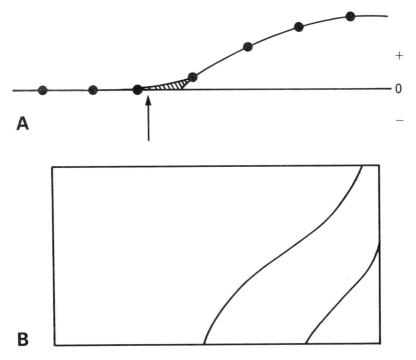

Figure 9.5 *(A)* Profile through isochore grid showing the effect of allowing the grid to take on zero values in the region where the mapped unit is absent. Dots represent grid intersections and values; arrow represents location of zero-line; shaded area represents mapping and integration error. *(B)* Map corresponding to *(A)*.

necessary. Table 9.1 lists procedures for controlling the location of a zero-line in thickness grids.

In Figure 9.6, a sandstone (unit B) overlies a shale (unit A). Both A and B are truncated by the unconformity surface C. The elevation of the unconformity is determined on surface C in all wells, but picks on A are made only where the stratigraphic top exists (chap. 5), so a Null value is entered for the top of A in wells where B is missing. We thus have a full set of structural picks on surface C, but some Nulls among the picks for the top of unit A.

Two different approaches exist for creating isochore grids: use the structural tops data, or calculate and use thickness values. The approach to use depends upon the geologic situation. For instance, the case depicted in Figure 9.6 shows thickness data that will be less likely to produce geologically reasonable results. In other situations, thickness data may be more appropriate than structural. Table 9.1 shows that the same basic methods are used, regardless of which data type is used. Four general procedures or methods are available for each data type, and each

TABLE 9.1
Possible Methods for Creating Isochore Grids

Degree of user influence	Thickness data	Structure data
None	*Method 1: Direct gridding of data*	
	Grid thickness data routinely	Grid structure data, subtract grids to give isochore
Moderate	*Method 2: Manual intervention at selected locations* Estimate values at selected points, combine with data, grid as in Method 1.	
	A Estimate thickness, honor the estimated values when gridded	A Estimate structure, honor estimated values when gridded
	B Estimate thickness ranges and constrain gridding to limits	B Estimate structure ranges and constrain gridding to limits
Low	*Method 3: Program-defined and digitized contours* Use presence-absence data and grid to define the zero-line, digitize line, combine with data, and grid as in Method 1	
	Use the (X,Y,0) values as thicknesses for gridding	Use the (X,Y) points along the digitized line to interpolate values from grid with control, combine with data from other surface, grid
High	*Method 4: Hand-drawn and digitized contours* Digitize hand-drawn zero-line, combine with data, and grid as in Method 1	
	Use the (X,Y,0) values as thicknesses for gridding	Use the (X,Y) points along the digitized line to interpolate values from grid with control, combine with data from other surface, grid

involves a different degree of manual intervention. We first summarize the two types of data and then discuss gridding methods.

The basic approach with structural data is to grid the fields individually to create two structure grids. One of these grids is subtracted from the other to provide a difference or isochore grid, as discussed in chapter 7. The methods described under structure data (Table 9.1) all involve this basic procedure.

The second approach — thickness-data methods — uses information in the struc-

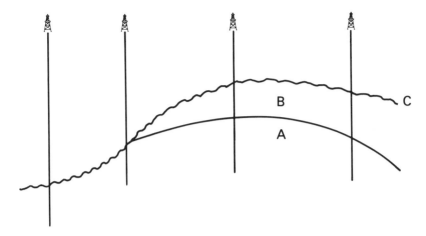

Figure 9.6 Cross section showing sand unit B above shale A, with both truncated by surface C. Structure picks are made on surface C in all wells, but on the top of A only where B exists (i.e., on the stratigraphic top of A).

tural tops for vertical wells to generate the isochore grid. Calculations on the input values in the two fields give thicknesses at each of the wells, with Nulls at those wells where sand unit B is missing. These data are then gridded directly to create an isochore grid. The methods described under thickness data (Table 9.1) all involve this basic procedure.

The four methods of manipulating data when creating isochore grids, regardless of whether the structure or thickness approach is used, may be summarized as follows.

1. *Direct gridding.* No special intervention is used, and the grids are made routinely.
2. *Manual intervention at selected locations.* Some manual intervention is done through interpretation of reasonable values to assign to those wells containing Nulls—that is, those wells in which sand unit B is missing. The estimated values are combined with the original data, and grids are created normally. A variant of this procedure uses the interpreted values as constraints (limits or ranges), rather than as values to be honored.
3. *Program-defined and digitized contours.* The program defines a reasonable location of the zero-line, based on presence or absence of the unit being mapped. This contour is digitized, the digitized points are combined with observed data, and the grid is created normally.
4. *Hand-drawn and digitized contours.* Manual intervention is made through hand-drawing the location of the zero-contour, digitizing that contour, combining the digitized points with observed data, and gridding.

187

Pitfalls of Erroneous Data-Collection Methods

Before we discuss the above procedures in detail, consider the impact of incorrectly recording and using data. First look at thicknesses. Assume we assign zero thickness values to those wells where sand unit B is missing, include the zeros with the observed positive thicknesses, and grid normally. This might seem reasonable, as zero is used as a data value where the unit has zero thickness (i.e., is absent). However, here zero has two different meanings: One is a point on a numerical continuum, whereas the other indicates the absence of a property. It is not correct to use a non-numeric (qualitative) property in conjunction with calculations involving quantitative thickness.

An immediate problem involves geological consistency. Figure 9.7A shows a thickness profile containing several wells and the result of using all these data points with a typical gridding algorithm. The wells penetrating the sand unit project a gradient toward the zero-line, and the grid honors the zero value at data point A, as it would all data points near the pinch-out. The zero-contour line will pass through all wells along the edge of the pinch-out region, implying that the sand unit goes to zero thickness precisely at each such well. While it might be

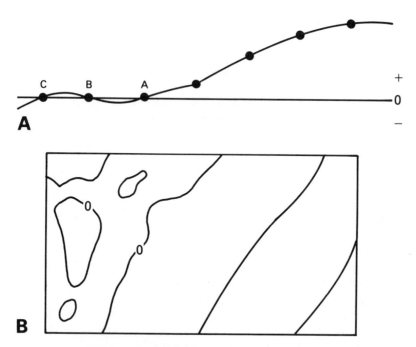

Figure 9.7 *(A)* Profile through isochore grid showing the effects of erroneously including zero values at wells A, B, and C where the sand unit is absent. The grid fluctuates about zero, rendering this method unacceptable. Dots represent thickness values at wells. *(B)* Map corresponding to *(A)*.

feasible for one well to have been drilled exactly on the edge of the unit, it is too great a coincidence to believe all wells were so drilled.

The thickness gradient in Figure 9.7A will still be influential past well A, so the grid will project somewhat negative and then be pulled back up to point B. It will project to a small positive value between B and C, and so on. Hence the grid, even though it honors all the data points, will fluctuate above and below zero.

Because this procedure gives unacceptable grid values in the region where the unit is missing, a contour map from that grid will also be unacceptable (Fig. 9.7B). Even if the amplitude of the fluctuations is less than the contour interval, the zero-contour will be erratic because it will honor each of the changes of sign of the Z-values. Many zero-contours will therefore "wander" through the missing region, while only a single zero-line may be appropriate on the map.

Figure 9.7A also shows that integrating this grid above zero will always result in a larger volume than is appropriate. The small positive volume between points B and C is incorrect but will add a positive value to the total. Only positive volumes are integrated, so the extra volume is not offset by the negative volume between points A and B.

Similar difficulties can arise with methods using structural data. Suppose unconformity tops are mixed with the correct picks for the top of unit A (Fig. 9.6). All wells will then have picks for both structural surfaces. In those wells in which unit B is missing, the two picks will have identical values. Grids based on the two data fields will honor the data points at these erroneously picked wells, and subtraction of the grids will result in a difference grid that fluctuates about zero, as does the grid in Figure 9.7.

In short, if data are not picked, recorded, and used according to stratigraphic principles, the resulting grids will be unrealistic. Quantitative and qualitative variables should not be mixed, and only stratigraphic tops are appropriate (chap. 5).

General Procedure

As stated previously, the two data types imply parallel procedures for each method. For thickness data, the method selected may require that additional information (e.g., digitized contour line) be combined with the original data. The resulting dataset is then gridded to create the isochore grid directly. Because zero-thickness points should not be used, they must be replaced by Nulls before gridding. Thickness methods are generally more appropriate for conformable intervals.

For structural data, the structure grids are created directly and then subtracted to provide the isochore grid. Again, depending on the method selected, additional information may be combined with the observed data before gridding. As before, data fields without a recorded value must have a Null assigned. This method is generally more appropriate for intervals crossing sequence boundaries.

Regardless of the method used, it is best if the gridding algorithm projects gradients beyond the edge of the data. If the algorithm extrapolates only local

189

averages, even extra information may not have great enough effect to cause the projection to define the zero-line.

The following subsections discuss the four methods presented in Table 9.1. For simplicity in each, we first describe techniques involving thickness data and follow with a discussion of structural procedures.

Method 1: Direct Gridding

Method 1 directly grids the observed data. This amounts to accepting whichever zero-line results from the data, regardless of potential problems. A difficulty with this method is that the zero-line may be poorly controlled and therefore appear to wander on the map. This is most likely to occur if only a few data points are located in the vicinity of the pinch-out.

When working with thickness data, complications can result if the observed thicknesses are highly variable. For example, the left side of Figure 9.8 shows thickness values that consistently decrease toward zero; this zero-line will likely be located in a reasonable position. On the other hand, if thicknesses increase and decrease along a profile, a single value can locally influence the zero-line to a great extent, possibly causing large irregularities. The right side of Figure 9.8 shows an extreme case of this, with the thickness values increasing toward the right; here a mapped zero-line would not result at all.

When working with structure data, two grids are subtracted to give an isochore grid. The zero-line will be defined where the node Z-values change sign, and it will be located by the projected gradients. Grids to be used for creating frameworks or for structural mapping would need modification by baselap or truncation procedures (chaps. 6 and 7). Do not use the modified grids to create the isochore grid because all grid Z-values will equal zero or Null in the region where the unit is missing, with problems as discussed above.

We recommend the direct gridding method for general use, at least as a first pass. The structure-type method is particularly recommended if several isochores in a complicated stratigraphic set are to be mapped. After the map is obtained, it should be evaluated to determine if the zero-line is acceptable. If it is, no further work is necessary. If not, then other methods in Table 9.1 should be considered.

Method 2: Manual Intervention at Selected Locations

If direct gridding (Method 1) does not give a good projection to the zero-line, manual intervention will be needed. The region around each well with a Null is uncontrolled, and the computer has no way to determine if the unit is present there or not. In Method 2, values are estimated geologically at each of the Null-containing wells in order to help control the grid and prevent positive projections. These estimated values are combined with the observed data, and gridding is done normally according to whichever data type is appropriate.

When working with thickness data, negative thicknesses are assigned to the

190

Figure 9.8 Profile showing thicknesses; the solid line represents the actual surface, and the dashed line represents the grid, and dots represent well-thickness data. On the left side, positive thicknesses project a consistent gradient to the well where the unit is missing (Null), resulting in a zero-line that is located reasonably. On the right, the thicknesses are erratic and there is no consistent projection to the zero-line.

wells in which the unit is missing. These negative values are obtained through geological interpretation of nearby data points and gradients and may involve hand-contouring the zero and some negative contours. The estimated negative thickness values are then combined with the observed positive thicknesses, and gridding is done normally. This method will control the region of the zero-line better because it will have information on the negative side.

A procedure for determining the negative values follows:

1. Make a base map, or post thickness values on the map from Method 1;
2. select two positive-thickness points near the pinch-out, such that a straight line from one to the other will intersect the zero-line as near perpendicularly as possible;
3. using proportional dividers or a similar tool, approximate the gradient between the points and project it toward and through the zero-line;
4. mark the map with the estimated location of the zero-contour, as well as several negative contours;
5. repeat Steps 2 through 4 for as many pairs of wells as possible (some points will be used in several pairs), but do not use pairs that give unrealistic or undesirable projections;
6. hand-contour the points marked in Step 4;
7. using the negative contours, estimate values at each of the wells where the unit is missing.

This process allows local changes in thickness gradient to exist, whereas other methods of projection might force a constant gradient everywhere.

The above procedure can be manpower-intensive and in many cases would overwork the problem. A simpler but often acceptable approach is to assign a single negative value to all data points containing Nulls. The constant negative value to be used is somewhat arbitrary and can be determined through experimentation. Common choices are the negative of the final map's contour interval,

191

the negative of the observed average thickness, or the negative of the nearest thickness value.

A similar process may be used with structure-type data. If one structural surface has few data points, its intersection with the other surface (which defines the zero-line) might be poorly controlled. Here we estimate values at those wells in which the stratigraphic top is missing and use them as control to modify the structural surface enough to give a good projection into the other surface. Working with structure grids is more difficult than working with thickness because it involves two complex surfaces, rather than only a single mapped thickness.

The procedure for determining the projected structure, as with thickness data, involves estimating values (elevation in this case) at wells with Nulls by the following steps:

1. Make a base map, or post structure values on the maps from Method 1;
2. select a few points in a critical region near the pinch-out and approximate structural values such that they are on the "wrong" side of the other surface — simple linear projections will not be appropriate here, as structural folding may be superimposed on the grids;
3. mark the map with the estimated value of each point;
4. repeat Steps 2 and 3 for as many sets of wells as possible (some points will be used several times), but do not use points that give unrealistic or undesirable projections;
5. hand-contour the points marked in Step 3;
6. using the hand-drawn contours, estimate values at each of the wells where the unit is missing.

As before, the thickness grid is created by subtraction of the structure grids. Because structural surfaces tend to be more irregular than isochores, this method may require more pseudo-points.

For both types of data, Method 2 has the advantage that the observed data are primarily responsible for the pinch-out, though it is augmented by the user who projects it into the missing region. The user's interpretation occurs mainly in the hand-drawn contours; however, these are used only for interpolation at the wells, so manual intervention is a minimum. The exact values used need only be able to extend gradients, so rough sketches are normally adequate. In addition, it is not necessary to be overly concerned with wells some distance away from the zero-line on the pinch-out side, particularly if several wells lie between the zero-line and the more distant wells. In this case, the distant points have a minimal impact on the zero-line, so their assigned values need only prevent the thickness grid from changing gradient and returning to positive values.

We recommend that values be estimated only at wells in which the unit is missing. This ensures that the density of data control will be similar over the map, matching the original control. If many additional pseudo-points are created, they

192

could dominate the observed data in the region of the zero-line; we wish to use actual observations as much as possible. Of course, if problems occur in areas with no well control, pseudo-points can certainly be created to fill gaps.

Table 9.1 lists the preceding as Method 2A. The closely related Method 2B results in slightly less manual influence. We may wish the grid to project in the general vicinity of the estimated points, but not necesarily to honor their values. Method 2B does not use the estimated values directly, but instead treats them as limits.

Chapter 5 describes procedures in which information in the form of limits can be incorporated into the gridding process. While discussed there in terms of structural surfaces, the same concepts apply to any variable, including thickness. Recall that the process consists of the following steps: (1) Separate the known values from the limits; the values of the limits indicate that the grids must be less than (BELOW) or greater than (ABOVE) the values at these locations; (2) grid using only known data points—that is, those that are not limits; (3) compare the grid to the limits; (4) if any limits are violated by the grid, combine them with the original data; (5) regrid with the modified data set; (6) repeat Steps 2 through 5 as needed.

For thickness data, we can use the estimated negative thicknesses as limits on the isochore grid rather than as values to honor. Here we specify that the grid must attain thicknesses less than—that is, BELOW—given estimates. This ensures that the grid will not project too shallowly or turn upward, but allows it freedom to project more negatively. A generalization in chapter 10 specifies tolerances on the estimated value (ABOVE and BELOW) and uses these to define thickness ranges within which the generated grid must pass.

We recommend Method 2A or 2B as effective ways to add control for gridding thicknesses. They do not overwhelm the program with artificial data, but do provide control for projecting the zero-line. They are also flexible, as additional information can be passed to the gridding program through increased numbers of pseudo-points.

Estimated structure-data can also be handled as ranges or limits. Recalling Figure 9.6, it seems likely that difficulties would occur if the structure grid on A projects too deeply. In this case we might use the estimated structural values to set limits rather than to be fully honored. If so, simply treating the estimated values as lower (ABOVE) limits would force the grid higher. Ranges might be defined for other circumstances, so the grid would be constrained by limits on both sides of the estimated value.

We also recommend Methods 2A or 2B for structure data, although these are more difficult to deal with than thickness data. Estimation of values can be a problem if the surfaces are structurally complex. In such a case the grids and data values could be referenced to a datum of the best controlled surface. In Figure 9.6, this would flatten the basal surface somewhat and simplify estimation. However, this operation essentially creates thickness, so we are back to the domain of Method

193

2: thickness data. In short, for complex structure or several intervals, the thickness methods are easier to use than the structure methods and are recommended for that reason.

Method 3: Program-Defined and Digitized Contours

This method specifies the exact location of the zero-contour primarily by use of the program and data. This line is determined by a grid that indicates presence or absence of the unit under study, with the user selecting a zero-line from the mapped grid. This line may be digitized and its points combined with the observed information for making the isochore map. Because of dense control along the line, the map will honor the selected line.

The first step in the procedure is to define the zero-line. For this, we treat the well information as qualitative (presence-absence) rather than as quantitative. Assign a data-value of 1 to each well in which the unit is defined — that is, where it has a recorded pick on both structural surfaces or a thickness value. Assign a value of 0 to all wells in which a Null appears in a structural field or thickness. A value of 1 thus represents "presence" of the unit in the well, and a value of 0 represents "absence" of the unit.

The zero-line is logically placed somewhere between wells with zero-values and those with one-values. In other words, if these values were to be contoured, the pinch-out would be placed where a contour has a value between 0 and 1. Create a grid using the presence-absence data and contour it; 0.1 is a convenient contour interval. Then select one of the contour lines as being a realistic location of the zero-line. Figure 9.9A shows an example of the presence-absence map.

The value 0.5 is usually selected as the zero-line unless other factors are involved. This value tends to place the zero-line midway between locations where the unit is known to be present and those locations where it is absent. Selecting a larger value (say, 0.7) moves the line closer to the presence wells, restricting the geographic extent of the unit. It also tends to steepen thickness gradients to the zero-line when later combined with the observed data. A smaller value (say, 0.3) enlarges the extent of the unit and creates lower gradients.

After the contour value is selected, the corresponding contour line must be digitized (as described in the first section of this chapter) so it can be incorporated into the data, or the procedure could be automated in the plotting program. When contours are drawn, most algorithms calculate the intersection of the given contour with the sides of each grid-square. If these (X,Y) locations are written to an output file, the contour line is virtually digitized. These points must then be incorporated with the observed data in order to create the isochore map. It is at this juncture that procedures diverge, depending on the method being used.

The case of thickness data is simple. Here the locations are treated as data, with each record containing (X,Y) coordinates and the value of the contour (e.g., 0.5). Simply change the Z-values to zero and combine these points with the original data. The combined thickness data are then gridded normally.

194

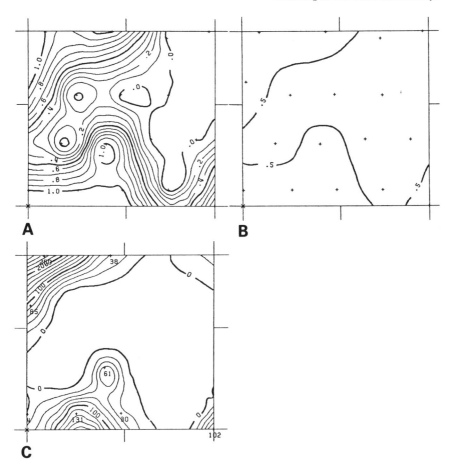

Figure 9.9 Maps showing steps in Method 3: Program-defined zero-line with presence-absence data. *(A)* Contours on the zero-one data. *(B)* The 0.5-contour. *(C)* Isochore grid generated with the 0.5-contour and thickness data. Note erroneous projected positive thickness in upper-right corner.

The case of structural data is a bit more complicated. Recall that the intersection of the two surfaces — that is, points with equal elevation — defines the zero-line. We therefore control the zero-line by forcing elevations in the two grids to be equal along the line, as follows: (1) Digitize the program-defined line; (2) using the better-controlled surface, interpolate a value from the grid at each digitized point; (3) combine these calculated values with the data from the poorly controlled surface; and (4) grid the combined data normally. This will result in a controlled structural surface, which can then be used to calculate the isochore grid through subtraction.

Regardless of whether thickness or structure data are used, the zero-line that

195

results from this procedure will be quite similar to the user-selected digitized contour. It may not be identical because the observed data can project gradients, but it will show the basic features.

Figure 9.9B shows the 0.5-contour from the presence-absence data, and Figure 9.9C shows the map generated by the thickness method. Study of the upper-right corner shows that positive thickness was placed in a location where the unit is absent; for this situation an estimated value, as from Method 2, would be needed.

A disadvantage of letting the program draw a potential zero-line with presence-absence data is that the line is dependent on data location. Good results will usually be obtained if data are evenly distributed, but clustered data can give an undesirable line. A second disadvantage involves thickness gradients; discussion of Method 4 points out that gradients can be controlled by digitizing several other contour lines, whereas Method 3 allows definition of only the zero-line.

Method 3 introduces a program-defined interpretation into the map and thus is a case of little manual intervention because the user's input consists only of selecting one of a set of potential lines. In spite of this, it is a powerful procedure. In some instances, however, it may be necessary to apply Method 2 in selected locations in association with Method 3.

Method 4: Hand-Drawn and Digitized Contours

This method introduces substantial intervention into the gridding process, as the exact location of the zero-contour is specified manually. Here a geologic interpretation of the exact location of the zero-line is made throughout the entire region, commonly in association with maps from Method 1. This line is drawn by hand on a map and digitized, as discussed previously. These digitized points are then combined with the original observed data and gridded.

Introduction of a digitized line is a straightforward process that should not cause difficulties. However, the user should be careful that digitized points are not too near observed data, particularly if a data point has a value different from the digitized contour. Close proximity will cause the gridded surface to make sharp bends in that neighborhood. If a value within one or two grid intervals of a contour is considered erroneous or untrustworthy, it may be best to delete the point and depend on the hand-drawn line completely.

Consider the case of thickness data. Digitizing only the zero-contour will not put a user-defined gradient into the grid and thickness might go to zero in an unacceptable way. Therefore it is often necessary to digitize and include additional (usually negative) contours. However, care must be taken. Drawing contour lines close together creates a map in which thickness increases rapidly from zero and then levels off to merge with the observed data values. On the other hand, spreading the contours could cause a too-flat gradient near the zero-line.

For structure data, recall that the intersection of the two surfaces defines the zero-line. We therefore control it by forcing elevations in the two grids to be equal along a user-specified line, as follows: (1) Sketch the desired zero-line on a map; (2)

196

digitize the sketched line—here we need only the (X,Y) locations of the points along the line; (3) using the well-controlled surface, interpolate a value from the grid at each digitized point along the line; (4) combine these calculated values with the data from the poorly controlled surface; and (5) grid the combined data normally. This process will result in a controlled structural surface, which can then be used to calculate the thickness grid through subtraction.

As with thickness data, we commonly need to digitize and include additional lines. This is a much more complicated process, however, because the added lines will not reflect where the surface has the same values, but where they have the desired difference in thickness. Suppose we wish to specify a line corresponding to the −10 contour. When we interpolate the value, we must add or subtract 10 from the structural values (depending on whether the well-controlled grid is above or below the other) before combining it with the observed structural data.

Even though digitized contours are used with both data types, the grid could reverse and come back through zero, becoming positive in areas where the unit is absent. Additional contour lines away from the zero-line may be needed to prevent uncontrolled behavior. These are generally far from the zero-line or data, so the exact digitized forms or values assigned are not important.

Method 4 introduces a specific interpretation into the map, and thus it is an extreme case of manual intervention. Positive contour lines could even be interpreted and used in gridding. This method has the advantage that difficulties with other methods can be overcome, but it requires careful usage. Interpretation of the contours should be done with care, and results must be analyzed to ensure that projections are as expected.

Additional Remarks

Table 9.1 summarizes various methods for creating isochore grids and maps. We generally recommend that Method 1 (structural data) be used first, as it relates to normal processing of stratigraphic frameworks. If the resulting zero-line is not acceptable, try Method 1 (thickness data), which takes advantage of the fact that thickness grids are sometimes better behaved than structure grids. Other approaches can be used if Method 1 does not work. Thickness-data procedures for Methods 2, 3, or 4 have greater ease of operation and interpretation than structural-data procedures.

This discussion has been concerned throughout with mapping thickness. However, similar considerations apply to other variables, such as the ratio of net reservoir thickness to gross thickness. The zero-line here can be based on the same considerations as mapping gross thickness. In addition, we also must deal with the contour line representing the limiting ratio of 100%. Here all the same principles apply, although now the grid should project to values greater than 100% in order to get a good 100% line. Thickness-based Methods 1 through 4 in Table 9.1 can be modified easily for use with nonisochore variables.

Some of the methods presented here (e.g., presence-absence and indicator gridding, gradient projection, and use of negative values) can be time-consuming and repetitious. However, their application is generally simple, so the methods should be amenable to programming. Programs would allow algorithms to perform some of these procedures directly without need for user intervention, thereby shifting the processing burden from a highly experienced computer user to a less computer-wise geologist. Such a shift does not imply that geologic interpretation can be ignored, but it would remove the need for routine data-handling considerations.

BIASED GRIDDING

Directional orientations or trends are commonly found in a geologic surface or variable. For instance, structural surfaces in western Pennsylvania contain strong trends due to parallel folding, and regional maps of Nevada show the north-south trend of the Basin and Range Province. Thickness maps can also show patterns, as depositional processes and sedimentation are influenced by hydraulic environment. Because such trends may be significant geologically, they should appear in contour maps.

The variable being mapped may contain trends, but the map may not show these trends if data density is low. Normal gridding algorithms operate on a nondirectional basis, trends not being preferentially introduced in the grid. The grids therefore tend to be isotropic, as an isolated data point may be surrounded by more-or-less circular contours. A contour map, while honoring the observed data, may not show the trends between data points.

This section discusses procedures whereby a directional bias or trend may be incorporated into a grid. Figure 9.10A shows a contour map that was generated normally, and Figure 9.10B shows a map from the same data, but with bias incorporated. Note that separate closures are joined, that contours are more linear, and that an obvious grain is present. However, all data values are honored in both maps.

The first step in using biased gridding is deciding if trends are present. At least three methods are available.

1. Base the decision on geological interpretation of a strong trend. The authors were involved in a project that required mapping uranium content in sedimentary rocks. Because the mineralization occurred in a braided-stream complex, the project geologist felt that predictability of grade was substantially better downstream than cross-stream. Maps prepared with a trend in the direction of paleo-flow represented the deposit better than did simple maps.
2. Use variograms or other geostatistical analysis to determine if directional continuity exists in the data. A set of directional variograms can be used to determine orientation and strength of trends. Calculation and interpreta-

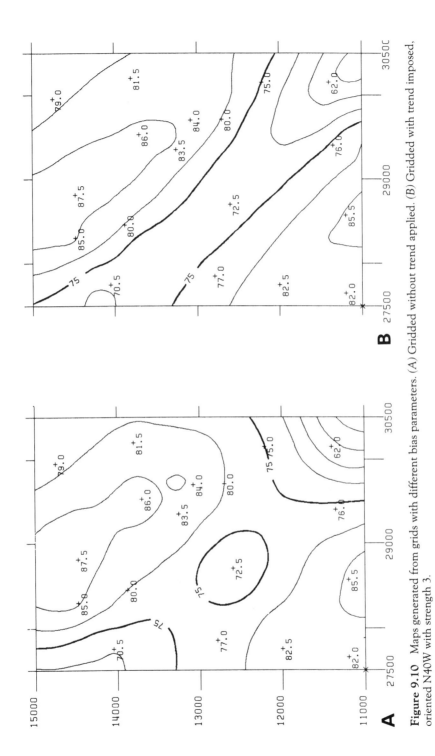

Figure 9.10 Maps generated from grids with different bias parameters. (A) Gridded without trend applied. (B) Gridded with trend imposed, oriented N40W with strength 3.

199

tion of variograms is discussed by David (1977), Journel and Huijbregts (1978), and Clark (1979).

3. Grid the data normally and contour the grid. If trends are apparent, consider regridding with bias to determine if map appearance is improved.

Other methods for determining whether or not to use biased gridding can probably be applied to specific projects, and there is no correct or incorrect approach. As with many geological techniques, each situation will likely require one or a combination of methods. The point to be remembered is that bias should be introduced only if there is a demonstrable reason for doing so.

Two pieces of information are required to define or specify the bias: direction or orientation, and strength or magnitude. The trend direction is conveniently specified by an azimuth from north. The strength or magnitude of the trend can be obtained by estimating how much predictability improves in the trend direction relative to the cross-trend direction. A ratio of 3.0 is a common magnitude. Figure 9.10B was generated with a direction of N40W and magnitude 3.

The direction and magnitude of bias can usually be estimated from previously generated maps of the surface. Either computer- or hand-drawn maps are used to determine the bias direction from aligned contour closures. An interesting but more time-consuming means of determining orientation is to select several different directions and, using a bias magnitude of 3, generate grids of that surface. Even grids with weak bias will tend to have fewer closures and smoother contours in the direction of trend than against it. Figure 9.11A-C shows the effect of three different bias directions with the same magnitude.

Bias magnitude is more easily determined from geologist's hand-drawn maps than from computer maps. With hand-drawn maps, measurements of length and width are made on several closures; the length divided by the width determines the magnitude. If the geologist's map shows a series of parallel ridges with essentially no closures, then select a large bias (4-6) to produce a similar effect. The same general approach can be used with computer-drawn maps; however, the ridges and valleys tend to be a series of aligned closures. Magnitude values obtained from measuring computer-generated closures should be increased by about half. Figures 9.10B and 9.11D compare magnitudes of 3 and 6.

Excessive time should not be spent in determining the exact magnitude and direction of bias. If the trend is strong enough to detect easily, then it probably should be used. Changing the direction by 10 degrees one way or the other, or changing the magnitude by 0.5 higher or lower, will have little impact on the resulting grid.

Biased methods can be programmed directly into gridding algorithms, although this is easier for a technique that calculates each grid node independently than for a spreading algorithm. A convenient procedure defines an ellipse oriented in the preferred direction, with the length of the major axis equal to the strength value and the minor axis having a length of one. The ellipse can be used to provide

200

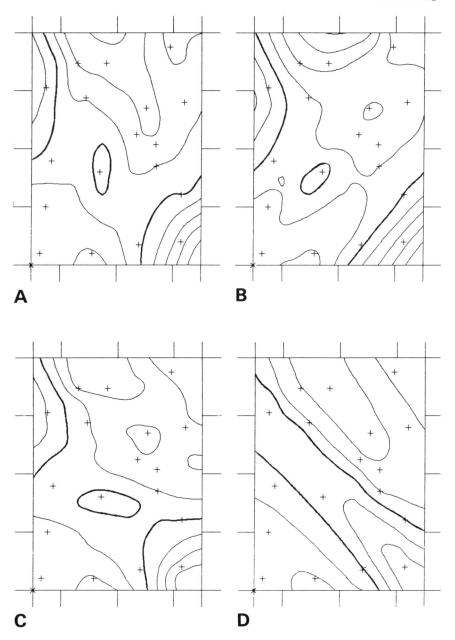

Figure 9.11 Comparison of different bias directions and magnitudes on data used in Figure 9.10. (*A*) Direction North, magnitude 3. (*B*) Direction N50E, magnitude 3. (*C*) Direction East, magnitude 3. (*D*) Direction N40W, magnitude 6.

weights for discounting distances in various directions between nodes and data points during grid calculation.

Because many programs do not include these automatic procedures, we present a series of steps that may be followed to generate a grid with specified bias direction and magnitude. These steps are illustrated in Figure 9.12, with three arbitrary data points indicated in Figure 9.12A. Specific program capabilities are required for some of these steps, as summarized in Appendix A.

1. Rotate the original data so that the bias direction is now pointing north (Fig. 9.12B). This is done with calculations on the X-Y coordinates in the original data file. If the program being used does not have rotation capability, the coordinates can be rotated through use of trigonometric calculations involving the angle θ between the bias direction and north. Rotation of coordinates (x',y') to (x,y) is done by

$$x = x' \cos \theta - y' \sin \theta$$
$$y = x' \sin \theta + y' \cos \theta$$

2. Compress the rotated data in the north-south direction by a factor equal to the bias magnitude (Fig. 9.12C). This is done by dividing the new Y coordinates by the magnitude and can be included with the rotation step.
3. Build an intermediate grid using the rotated, compressed data (Fig. 9.12D). The grid interval should equal the final desired grid interval divided by the bias magnitude, so the grid interval will be smaller than the final interval.
4. Convert the intermediate grid to data by use of the grid-to-data task. The grid is thus converted to a number of (X,Y,Z) records in a data file. These are expanded in the north-south direction by multiplication of the Y coordinates by the bias magnitude (Fig. 9.12E). This expansion is the reverse process of Step 2.
5. Rotate the expanded grid-data back to the original bias direction (Fig. 9.12F). This is the reverse process of Step 1.
6. Combine the original dataset with the expanded and rotated grid-data. Any grid-data points that are near an original data point are removed from consideration (Fig. 9.12G), thereby ensuring that original data values will be honored without local interference from the generated grid-data points. A distance limit of 1.5 times the final grid interval has been found reasonable for deciding if a grid-data point is too close.
7. Grid the combined original data and edited grid-data; use the desired grid interval when producing the final biased grid (Fig. 9.12H).

These steps were used to generate the biased grids shown in Figures 9.10 and 9.11. The procedure seems complex, but no step is difficult.

Although most projects will not require biased gridding, it is a useful tool to have available. We recommend that a general set of program controls be con-

202

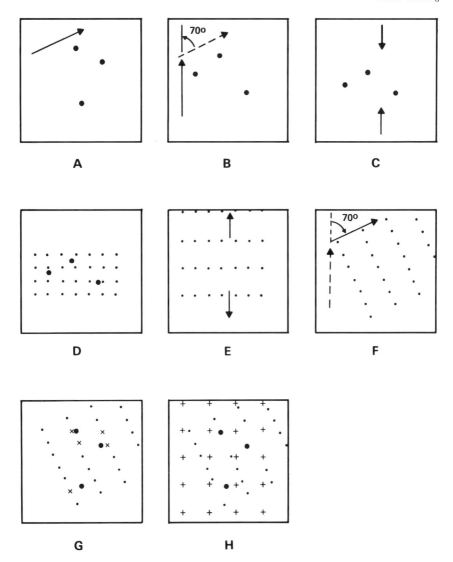

Figure 9.12 Steps in the biasing process. (A) Map of the original data. (B) Rotate the data to orient the bias direction north-south. (C) Compress the rotated data in the north-south direction by the bias magnitude. (D) Grid the compressed data. (E) Expand the grid in the north-south direction by the bias magnitude. (F) Rotate the expanded "grid-data" back to the original bias direction. (G) Eliminate the "grid-data" too close to the original data. (H) Grid the combined data and remaining "grid-data" points.

203

structed to perform these steps. The controls can then be saved for later use as a "template job." Whenever biased gridding is needed, the template can be modified slightly for use with the current project. Some programs allow "macros," or groups of operations, to be defined; biased gridding is useful in such a form.

It should be pointed out that the bias operation will have little effect with high data density. If only a few nodes occur between data points, then honoring the data allows little flexibility for putting in trends. In this case, any trend in the data will show up automatically from a normally created grid.

CHAPTER 10

Selected Petroleum Applications

Detailed mapping is necessary for petroleum exploration and production; accordingly, many computer-mapping applications exist in this field. Much exploration mapping deals with structural features (i.e., large folds or faults) or stratigraphic relationships, both of which have been discussed previously. The applications covered in this chapter are more closely related to petroleum production than to exploration.

FLUID CONTACTS

Creating Grids

Contacts between different fluids are important for defining the limits of a reservoir. The oil portion of a reservoir is the interval above the oil-water contact (OWC) and below the gas-oil contact (GOC). Knowledge of these surfaces and the structural top and base of the porous reservoir unit are required to define the thickness and lateral extent of oil in the unit (cf. Fig. 2.11). Similar considerations are necessary to map the gas interval.

The first step in mapping the OWC is to obtain data on its depth. This information normally comes from interpretation of well logs because the physical properties of water-bearing rock differ from those of oil-bearing rock, and an experienced log analyst can usually detect the change. A less common method of determining the OWC is through analysis of water-saturation measurements in cores.

After the depths of the OWC are determined in the wells and corrected to elevation, several methods are available for creating a grid. The simplest assumes that the OWC is a horizontal plane. The elevation of the plane is normally given by the average of the observed OWC elevations. In this case, each grid intersection is merely assigned the average value. This is usually a good approximation unless potentiometric gradients or variations in permeability exist (Berg, 1975).

If the OWC is not flat, a dipping plane may be more appropriate because it allows hydrodynamic tilt or the effects of permeability trends to be modeled. This plane is normally defined by a least-squares fit (chap. 11). Other low-order trend functions (e.g., quadratic) might be used if a plane is judged too simple. However, as with the first method, the trend surface will honor only the general tendency of the data, but not the exact values.

The OWC is sometimes mapped by hand. The geologist or engineer interprets the surface and draws the contours smoothly. If only an OWC map is required, there is no need for the computer. However, for interaction with other computer-generated surfaces, the OWC must be put into grid form by digitizing the contours and then using the generated points to create a grid (chap. 9).

Finally, the OWC grid can be created simply by gridding the observed data, exactly as is done with any individual structural surface (chap. 4). Recall that some of the data for mapping a structural surface may consist of limits (e.g., the surface lies below the bottom of the hole), as discussed in chapter 5. An additional complication can arise here, as the OWC values may include ranges of allowable elevation as well as limits, discussed below.

Similar procedures are followed for mapping the gas-oil contact. In some reservoirs, tar lying between water and oil constitutes a fourth fluid. Tar-oil and water-tar contact grids can also be made with the same methods. Tar contacts will not be discussed here, but the same general procedures apply to them.

If two or more fluid contacts are to be mapped, special care should be taken to ensure that the surfaces are correctly related to each other. As with stratigraphic horizons, the OWC and GOC cannot cross and are usually treated conformably. The methods discussed in chapter 6 for conformable surfaces are followed: (1) Map one of the surfaces (say, OWC) directly; (2) map oil thickness; and (3) combine the OWC structure and thickness grids to create the gas-oil contact grid (GOC). If the fluid goes to zero thickness (pinches out), considerations discussed in chapter 9 for zero-lines are appropriate. A second method that could be used for sharply intersecting contacts is to grid the two surfaces directly and then relate them through truncation or baselap (chap. 6).

Data That Represent Ranges

In the last section of chapter 5, we discuss data points that cannot be used directly to create a grid but that nevertheless contain information that must be honored or taken into account. For instance, a hole may not be deep enough to

206

penetrate the surface, so it cannot be used for gridding, but the lack of data indicates that the surface cannot be shallower than the bottom of the hole.

To generalize from data limits to ranges, consider the case of mapping fluid contacts. Figure 10.1 shows a cross section through a sandstone reservoir bounded above and below by shale. An oil-water contact surface, OWC, passes through the reservoir but projects into shale outside of the productive unit. Wells 3 and 5 have picks that can be used to grid the OWC surface; these are called fixed picks.

Unfortunately, none of the other wells in the figure have usable picks. Well 2 has a total depth (TD) that is too shallow, so it should be treated as a limit when generating the OWC grid, as described in chapter 5. We record two items for this well: the elevation of TD, and a BELOW code; these indicate that the OWC is deeper than, or below, the recorded value.

Similar restrictions are found in wells 1 and 6 where the OWC intersects impermeable shale and could not be picked in the wells. In well 1 the sandstone is oil-filled, so we know only that the projection of the OWC is deeper than the base of the reservoir at that location. In well 1, therefore, we record a BELOW code and the elevation of the base of the reservoir. Well 6 is the opposite case. It penetrates the reservoir in water, so the OWC must be shallower than, or above, the top of the sandstone. In this case, an ABOVE code and the elevation of the top is recorded. We do not claim that the OWC exists in the shale, but that its projection would be located there.

Well 4 shows a more complicated situation. Here an internal shale zone is intersected by the OWC, so the OWC cannot be detected in the well. In this case we

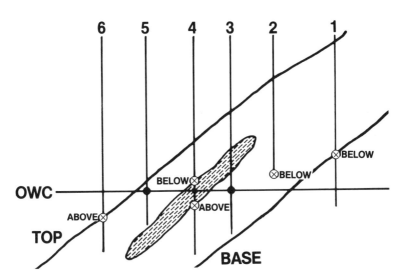

Figure 10.1 Types of oil-water contact data. The depth to the contact is known exactly in wells 3 and 5, but is known only within a limit in wells 1, 2, and 6, and within a range in well 4.

207

know only that the depth to the OWC is below the top of the shale body and above the base of the shale. This defines a range—merely a case of two simultaneous limits—within which the OWC is located. The data record must therefore contain two elevation values (the top and base of the shale), plus an indication of the range.

When the data values are submitted to the computer, fixed picks are entered in a field normally. The picks defining ranges or limits must also be entered in two fields, although one of them can be the same field that contains the fixed points. A third field is necessary to code the data point as ABOVE or BELOW; a range can be recognized by the code RANGE or simply by the presence of elevation values in two fields, and a fixed pick is designated by a blank in the code field.

Table 10.1 shows recorded data for the six wells in Figure 10.1. Here the (X,Y) coordinates are not listed, but they would normally be included. Field Z contains the fixed picks and limit values, while field ZR contains elevations of the deeper range pick. Field CODE indicates the type of data point. A brief comment is also included, although it would not be put in the data records.

The procedure described in detail in chapter 5 is used. However, now tests for each side of the range, rather than only a single limit, are made. Begin by using only the fixed picks to create a grid. Compare this grid to the limits and ranges. If a constraint in a well is honored, there is no problem. However, if constraints are violated, convert the limits or ranges to fixed picks (use the end of the range nearest the violating grid value), combine them with the others, and regenerate the grid. Then test the new grid against the remaining constraints.

This is an iterative process; one or two iterations are usually adequate. The reason for additional passes is that correction for a constraint may alter the grid enough that other nearby constraints are now violated. Additional iterations are more likely to be needed for ranges than for only single-sided limits.

MAPPING GROSS ROCK THICKNESS

A commonly used map shows gross rock interval thickness associated with some fluid. For example, suppose we are dealing with gas and wish to know the

TABLE 10.1
Example of Input for Data Containing Limits and Ranges

X	Y	WID	Z	ZR	CODE	
—	—	1	−9024		BELOW	base reservoir elevation
—	—	2	−9095		BELOW	TD elevation
—	—	3	−9133			fixed pick
—	—	4	−9110	−9162	RANGE	range
—	—	5	−9137			fixed pick
—	—	6	−9205		ABOVE	top reservoir elevation

thickness of rock in the reservoir that corresponds to the gas interval. In Figure 10.2A, the gas interval is defined above by the top of the reservoir unit, T, and below by the GOC; the hatchured portion indicates gross gas.

When creating thickness grids through grid manipulation, we have found the concept of an envelope to be useful — that is, finding one surface that limits the top and a second surface that limits the base of the interval of interest, so that the two surfaces form a surrounding envelope. In Figure 10.2A the envelope is defined by

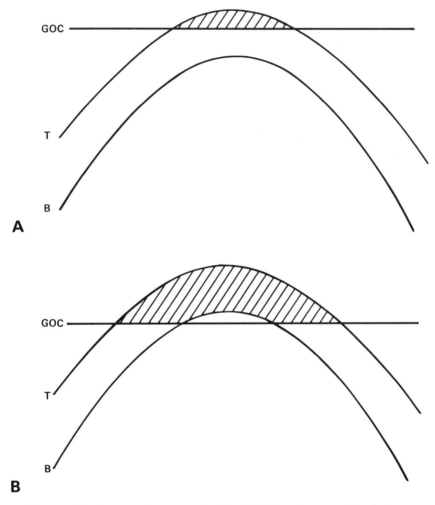

Figure 10.2 Gas interval in a reservoir. T and B indicate the top and base horizons, GOC indicates the gas-oil contact, and the hatchured area indicates the gas interval. (A) Gas bounded below by the GOC. (B) Gas bounded below by the combined GOC and horizon B.

209

the reservoir top and the GOC, and the thickness grid is obtained by subtracting the base of the envelope from the top.

Figure 10.2*B* also shows gross gas, but here it is limited below by the combination of the GOC and the base of the reservoir, B. In this more general case, we are dealing with three surfaces, two of which are stratigraphic. The GOC grid can cross the T and B grids. However, we use the concepts and operations discussed in chapter 6 to combine and relate the various surfaces in order to create the envelope grids.

The top of the envelope is simply grid T. However, the base is a combination of B and GOC. No gas in the reservoir can occur deeper than either B or GOC. We therefore operate the two grids to output the shallower (greater elevation) of the two corresponding values at each grid intersection. In stratigraphic terms, we lap GOC onto B to create grid BB. Simply subtracting the grids as T − BB provides the grid of gross gas thickness.

The gross thickness in the central part of the reservoir of Figure 10.2*B* equals the total reservoir thickness, and gross thickness decreases at the edges. Note that subtraction of the two grids allows negative values outside the hatchured portion of Figure 10.2*B*. As pointed out in chapter 9, the grid should be allowed to project negatively in order to plot better zero-lines and to increase the accuracy of volume integrations.

Turning now to oil, the hatchured portion of Figure 10.3 indicates the gross oil interval. Here both the top and the base of the envelope are combinations, as we are dealing with four surfaces. The basal limit is similar to that described above; grid BB is created by lapping OWC onto B. The upper limit of the envelope is defined by grid TT, which consists of the deeper of the two grids because oil cannot be any shallower than either the GOC or the reservoir top, T. Truncating T by GOC gives grid TT. Subtraction, TT − BB, gives gross oil thickness. Again, the thickness grid should be allowed to take on negative values.

An advantage of this procedure is that a well-defined wedge zone is obtained. In the central reservoir of Figure 10.4, gross oil thickness is equal to that of the reservoir. However, at the outer edge of the reservoir, a wedge is created where gross oil is confined by the OWC and the dipping formation. The projected surfaces give a good definition of this thickness. A small error at the inner wedge at the base of the interval can result. This error, generally not detectable in maps but possibly affecting integrated volumes, is discussed in the volumetrics section.

The following alternative method is incorrect: to measure and record, at each well, the actual gross oil thickness (where oil is absent, record a Null value) and grid these thicknesses normally. The map will honor the data at each well, but the wedge will not be defined between the wells, and the abrupt thickness change will be lost. Even special handling of the zero-line will not cure this problem. Difficulties with this procedure are increased if complicated stratigraphic relationships exist between T and B. This simple method therefore should be avoided (cf. Wharton, 1948).

210

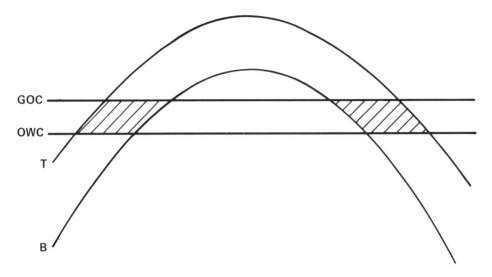

Figure 10.3 Oil interval in a reservoir. T and B indicate the top and base horizons, GOC indicates the gas-oil contact, OWC indicates the oil-water contact, and the hatchured area indicates the oil interval.

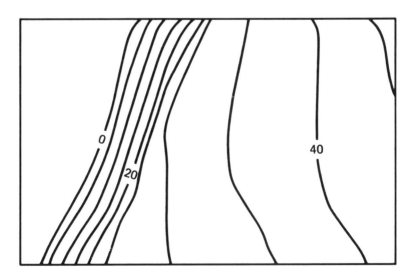

Figure 10.4 Contour map of gross oil thickness, with well-defined wedge indicated by the closely spaced contours. (After Jones, 1984, p. 661)

211

VOLUMETRICS

Knowledge of the petroleum reserves or volumes in a field is extremely important. The economics of further drilling or development is impacted heavily by the total volumes of the fluids present. Planning is similarly affected by the geographic distribution of the fluids. In addition, if the field is operated as a unit with several owners, the relative equity positions of the owners must be determined. *Volumetrics* is the term used to describe the calculation of these fluid volumes.

General Procedure

The goal of the volumetrics-calculation process is to compute the reservoir volumes of the hydrocarbon fluids: gas, oil, and possibly tar. The volume of fluid in a trap depends on the size of the reservoir (thickness and lateral extent), the amount of available pore space to hold fluids, and the proportion of water in the hydrocarbon fluid.

It is necessary to calculate the hydrocarbon pore volume (HPV) of each fluid. This is normally obtained by first calculating hydrocarbon pore thickness (HPT) at many locations by the product

HPT = gross hydrocarbon thickness × ratio of net to gross thickness
× average porosity × hydrocarbon saturation.

The HPV is computed by integration over the HPT locations, normally with separate calculations by lease and fault block, to give total oil or gas in place.

Table 10.2 defines the terms in the equation. Each of the four components in the formula can be calculated at many locations and mapped for use in the formula. The formula can thus be applied at a single location or, more usefully, over an entire fault block or reservoir. The terms in the general formula are multiplied in order, giving partial products of gross rock thickness (GT), net rock thickness (NT), pore thickness (PT), and finally HPT for the fluid in question.

In pre-computer days, these calculations were made manually. Hand-drawn maps of the several variables were combined, generally through cross contouring, to create a map of HPT. This map was then manually integrated by use of a planimeter to provide the required volume, HPV. This tedious method was commonly simplified by assuming constant values of porosity and water saturation throughout and calculating net thickness directly. With the computer, however, we create grids of the variables and then form the product HPT as a grid. Numerical integration, discussed below, provides HPV.

Most of the variables used in the general equation may be assumed to be continuous fieldwide, with one possible exception; if faults exist, the structure grids on the top and at the base of the interval are discontinuous. Such a variable as

TABLE 10.2
General Volumetrics Formula

---

hydrocarbon pore thickness = gross hydrocarbon thickness ×
ratio of net to gross thickness ×
average porosity ×
hydrocarbon saturation

where

gross hydrocarbon thickness is the thickness of the hydrocarbon within the interval

ratio of net to gross is the ratio of thicknesses of porous (pay) rock in the gross interval under consideration

average porosity is the average of that porosity making up the porous part of the net/gross ratio calculation

hydrocarbon saturation is the fraction of the fluid that is hydrocarbon, calculated only for the hydrocarbon column

hydrocarbon pore thickness (HPT) is the thickness of the actual hydrocarbon at a location

hydrocarbon pore volume (HPV) is the total volume of fluid hydrocarbon within the interval, integrated over the area A, as

$$HPV = \int_A HPT \, dA$$

---

net/gross ratio normally would not be influenced by faulting unless deposition was contemporaneous with faulting. Similarly, porosity would be continuous fieldwide unless severe faulting with substantially different depths of burial caused different degrees of porosity reduction. Adjustments in the procedure caused by faulted structure are discussed below.

In applying the general equation, first consider the geometric framework of the reservoir. The top and base of the reservoir and the fluid contacts define the vertical limits of the fluid and thus the gross rock thickness (GT) for the fluid in question. Procedures discussed in the previous section for creating gross-thickness grids are followed here to define the "container" for the hydrocarbon. Integration of thickness provides gross rock volume in the oil interval. Similar methods are used to calculate gross gas thickness.

The second step defines net fluid thickness (NT). Any typical interval includes nonproductive shale or nonporous or impermeable zones that reduce the thickness of the productive reservoir. The effect of these zones must be removed from the gross interval. On a detailed basis (say, foot by foot) at each well in the reservoir, the geologist must determine whether "pay" (that is, porous, productive, and so on) rock or "nonpay" rock is represented, and the total thickness of each is measured. Of course, the term *pay* represents potential pay, as determinations are

213

made without regard to whether a hydrocarbon fluid actually exists in all or part of the gross interval. Before net pay is calculated, thicknesses in deviated wells must be converted to a vertical reference (cf. Pennebaker, 1972).

Why not just use the net-thickness grid directly and skip the GT step? The answer is that such features as the wedge zone will not be defined properly, particularly if the structure grids were created by more data points than are available for pay determination. This fact has long been known (Wharton, 1948) but is often ignored.

At this point the user must make a choice, as the net/gross ratio grid may be generated in one of two ways: (1) Calculate the ratio of net pay thickness to gross interval thickness for each well, and then grid and map the ratios; or (2) grid the thickness of pay rock, and divide this grid by the gross-thickness grid to obtain the ratio grid. Whether Method 1 or 2 is better depends on the relative variations of pay and gross thickness, but Method 1 is used more commonly.

Method 2 can have complications if pay thickness is near or equal to gross thickness. In this case, projections of the net thickness grid could exceed gross thickness, leading to inconsistencies. Special care must be taken to correct these projections, as well as to account for the 100% line. If nonpay is generally thicker than pay, gridding nonpay and subtracting the nonpay grid from the gross thickness grid may minimize error in pay thickness.

Regardless of how it is obtained, net/gross ratio attains a value between zero and one and represents the proportion of the gross interval that contains pay. The net rock thickness (NT) grid for this fluid is determined by multiplying the gross fluid thickness (GT) grid by the net/gross ratio grid. This grid can be mapped to show geographic distribution and integrated to give net rock volume corresponding to the fluid in the interval. As before, volumes can be restricted to specific leases or fault blocks.

A similar procedure is followed for the third step, porosity thickness (PT). PT measures the volume of the original "container" that is available to contain fluids. At a given location, it is the total pore space, combined into a single hole. Average porosity corresponding to the rock designated as pay is calculated in the interval used for determination of pay; data in deviated wells should first be corrected to a vertical reference. This information is gridded, mapped, and passed to the volumetrics process. Multiplication of the average porosity grid by the net-thickness grid (NT) gives a grid of porosity thickness (PT). Integrating PT gives pore volume for the interval and fluid.

The fourth step in the procedure requires a grid of hydrocarbon saturation. Recall that water is found within the oil column above the OWC, the amount increasing as the OWC is approached. Two methods are available for calculating a grid of oil saturation. The first is to calculate or measure oil saturation at each well and then grid these values. This method ignores the effects of lateral permeability variation and varying height above the OWC, although it may give a good approximation if the transition zone is thin. The second method, generally more accurate,

214

uses empirical or engineering functions of height above OWC, porosity, permeability, or other variables. Special programs would be needed to read several grids and apply the formulas at each intersection to make a grid of hydrocarbon saturation.

Multiplication of the porosity-thickness grid (PT) by the saturation grid gives hydrocarbon pore thickness (HPT). HPT is the amount of hydrocarbon fluid at each grid intersection; if PT represents the height of a column of fluid in a tank at a node, then HPT specifies the amount of fluid that is hydrocarbon at the node in the tank. HPT is integrated to give hydrocarbon pore volume (HPV)—that is, the volume of hydrocarbon fluid.

The general volumetrics procedure that we have outlined has two major advantages. First, data on the several variables come from different sources. For instance, structural information may be derived from well tops, seismic interpretation, or hand-drawn maps. Net/gross ratio information is normally available at most wells, while porosity may be restricted to wells with cores or specific petrophysical logs. Hydrocarbon saturation may be calculated by separate equations. If a volumetrics procedure requires all variables to exist at each data point, the amount of usable information would be reduced severely, and some information would have to be wasted.

Second, the modular process consists of a number of separate steps, each of which can be handled independently. Difficulties in one variable can be solved without need to involve the others. Maps and volumes can be obtained at each stage, allowing stepwise checks to be made. In addition, average reservoir values can be obtained (e.g., porosity averaged over the reservoir and weighted by net rock volume is found by dividing integrated PT by integrated NT). Finally, once the individual variable grids are accepted, calculation is a noninterpretive, mechanical process.

Numerical Integration

Virtually all mapping programs contain an option to integrate a grid numerically. This integration is simple conceptually, but many program refinements can be made in order to gain accuracy and precision. The common method of numerical integration is based on the definition of a double integral. Figure 10.5A shows a portion of a grid. A prism extending from the integration datum (zero) to the surface is drawn at a grid-square or -rectangle. The volume of that prism, given by $L \times W \times H$, is the value of the integral over that part of the grid, where L and W are the length and width of the mesh and H is the average height of the surface.

Because H is not known (it itself is an integral), we must make further calculations. The value of the surface is known at the four intersections, so they could be averaged to estimate H. However, this calculation may not be accurate if the surface is curved in this region. Division of the block into smaller pieces improves accuracy because it creates two (Fig. 10.5B) or more (Fig. 10.5C) triangles; the volume of

215

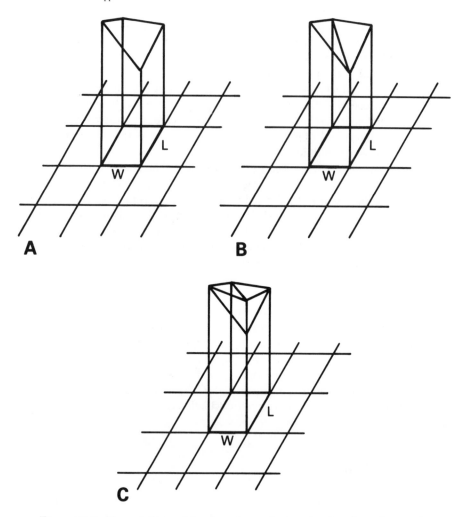

Figure 10.5 Numerical integration through use of prism placed on the grid-rectangle of dimension $L \times W$. The top of the prism is defined by the gridded surface. (A) Single prism used to compute volume $L \times W \times H$, where H is the average of the four node values. (B) Prism divided into two triangular prisms to gain accuracy. (C) Prism divided into four triangular prisms.

each piece is then computed by multiplying the triangular area by the average height. However, now one of the vertices in Figure 10.5C does not correspond to a grid intersection, so its height is unknown and must be estimated.

Estimation can be done by simple interpolations, splines, or low-order functions as with inverse interpolation (Appendix A). With the heights estimated and curvature included by use of additional intersection values, the volumes of the small triangles can be calculated and summed to give the estimated volume. The same procedure can be followed for all portions of the HPT grid, giving HPV.

If the entire grid is to be integrated without boundaries, the process is relatively simple. However, in practice we must deal with edges—that is, it is necessary to integrate up to a lease boundary, the edges of a fault block, or an irregular zero-thickness contour. In most cases, the leases and faults are defined by polygons while the zero-line is defined by grid projection or a digitized line, and calculations must extend exactly to the boundary. These complications can be significant and are encountered in most mapping projects, so the integration procedures must handle them. Such complexities have occupied programmers for years.

Let's return to the concept that *H* represents the average height of the surface in a portion of the grid. Rather than approaching the integration as summing volumes of prisms, we can estimate *H*. A common numerical procedure is to select many points in a simple gridlike pattern, determine the value at each point, and average them as an estimate of *H*. If only the four grid values are available, a simple average may not be representative. However, intermediate locations can be estimated by using inverse interpolation with the surrounding grid intersections, as discussed above. Many (but not too many) points are estimated and averaged to give *H*, which is then used to calculate the volume.

This method appears to differ from that previously discussed. However, if each point is considered as representing the top of a tiny square or rectangular prism, then we again have the basic method. As before, boundaries must be considered, and this may be more difficult to do exactly. In this case, only points within the valid portion are used, and points falling outside are skipped.

The previous discussion has been in terms of integrating a grid of HPT. However, the same process can be followed if a grid equivalent is used. Recall from chapter 4 that this is essentially a dataset containing several variables, each corresponding to a grid. The general equation described above is used to combine each of these variables to give HPT at each of the calculation points in the dataset. As with a grid, values at the vertices of the estimation triangles can be calculated by interpolation methods, or numerous values from the surface can be used to estimate *H*.

The grid-equivalent method has an advantage over a grid. Intermediate grids often contain abrupt changes in slope (e.g., at the wedge zone), and numerical integration methods will not accurately represent sudden changes in the surface. Each grid-equivalent point contains all the variables (structure, thickness, porosity, and so on), and these individual values will be smoother locally than will the product. The order of processing is reversed, as the integration triangles are determined for each separate variable first. The values at each vertex are then combined through the general equation, and the combination is used with the triangular area to give volumes. This procedure can help accuracy, although complications exist at the zero-line.

This section has described a generalized volumetrics procedure for calculation of total original hydrocarbon in place. The following section expands the general

217

discussion to include special considerations that must be taken into account in any given project.

Special Considerations

Subdivision of the Reservoir

The general volumetrics procedure discussed here assumes that geologic variations (net/gross ratio or porosity) are not significant vertically. At a location, the general equation applies a single value of net/gross ratio or porosity to the entire interval. If there should be variation (e.g., if porosity is greater at the upper part of the reservoir than at the base), the assumption would not be valid.

Even if there is little geologic variation, hydrocarbon saturation will not be constant with depth. Recall that water saturation varies as a function of height above the OWC, being greatest in the transition zone near the OWC. Near the OWC, average hydrocarbon saturation might not be representative of the bulk of the interval. Further, the region of rapid change in one variable (e.g., porosity) might correlate with that of another (e.g., hydrocarbon saturation).

Although the general volumetrics procedure could be applied to the entire, thick reservoir interval, accuracy and precision are gained if the calculation interval is thinner. Accordingly, a common practice is to divide the reservoir into zones by use of log correlations and geological interpretation. If the reservoir is thick or heterogeneous, zones defined originally for such other purposes as correlation might be used.

If the zones were not originally included, they must be fit into the existing framework by adding horizons within the existing sequences. These operations are done in conjunction with pretruncated or prebaselapped grids, and the stratigraphic relationships are applied at the end.

In some situations (for instance, multi-owner equity determinations), the total reservoir-thickness grid may be created and agreed to be correct. When zones are defined later, we would develop a set of zone isochore grids. We could add these grids together to obtain a total-thickness grid and compare it to the agreed-upon total-thickness.

The two grids purport to measure the same total thickness, but it is likely that they will differ. Differences away from data points may not be too surprising, but it is also possible to find disagreement at a well even if the program honors all the data, if some zone picks are not made for one of several reasons. In this case, thicknesses in certain zones or the total reservoir may be based on estimates rather than observation.

Data points should be corrected where they disagree. Methods for making this correction include the following steps: (1) Determine the total thickness at each well through inverse interpolation; (2) determine zone thicknesses through inverse interpolation on the zone-thickness grids; (3) at each well, determine the cumulative error; (4) where discrepancies exist, distribute the error on a proportional basis into those zones where no observed thicknesses were available; (5) regrid the altered dataset. Steps 3 and 4 might require the use of an external program, as typical mapping packages do not include the capability to do these percent corrections. Similarly, percent-correct the zone-thickness grids directly on a node-by-node basis.

Even these correlated zones might be considered too thick, and further subdivision might be required. These subdivisions (subzones) might have no geological or petrophysical basis, but would be defined for numerical purposes only. However, some geological concepts must still be taken into account. For instance, if the zone was formed by onlapping sedimentation, the subzones should onlap onto the surface below. These subzones are created by subtracting constant-thickness isochore grids from the top of the zone. Similarly for truncation, constant-thickness subzones are added to the base of a zone. For a conformable zone, all subzones are identical and of variable thickness. As with zones, the subzones should be incorporated into the stratigraphic framework to be compatible with the sequences and their relationships.

The volumetrics procedure described above with the general equation can be applied to every subzone or zone: Generate grids for all subzones for each fluid, and apply the equation to calculate the various fluid-thickness grids. It is a simple matter to sum the subzone or zone isochores into a total reservoir isochore and then integrate this total grid. It is also possible to form separate zone grids at the same time.

Faulted Reservoirs

As was pointed out previously, the variables in the general equation can be gridded fieldwide; porosity, net/gross ratio, saturation, and reservoir-thickness grids can be assumed to be continuous. However, structure grids might be discontinuous because of faulting. The methods described here assume vertical faulting, which simplifies matters considerably. Analysis of slanted faults is substantially more complicated because of fault gaps. However, extension to slanted faults follows directly from the methods discussed in chapter 8.

Two general methods of using faulted structural surfaces exist: using a single fieldwide structure grid and using separate structure grids for each fault block. The single-grid approach incorporates the faults into a continuous grid. In this case, a single structure grid is generated that contains abrupt offsets in elevation at the faults. This grid could be used fieldwide, although calculation errors in the vicinities of the faults can result if integration is not done properly.

Three choices exist for creating fieldwide structure grids. Probably the most common involves digitizing a hand-drawn structure map, as discussed in chapter 9, and then gridding the digitized data. The continuous grid can be created through use of the restored-surface or the fault-trace method (chap. 8).

Errors can result with use of fieldwide continuous grids at the integration stage.

219

Most integration programs use from 4 to 16 intersections in the vicinity of the location being calculated to estimate the surface with such smooth functions as polynomials or splines. However, if a fault cuts through these intersections, the function will give a poor representation because of the abrupt change in elevation. Volume will be too low on one side of the fault and too high on the other side.

To remedy this problem with fieldwide grids, the integration program must pay special attention to the faults. The edge of the fault block must be detected, and only intersections within the block being calculated should be used. Mapping programs that use the fault-trace technique to build a continuous grid usually contain a similar procedure for volumetrics that detects the edge of the fault block and restricts calculations.

Most volumetrics processing uses the fault-block polygon method. This results in a different structure grid for each fault block, with the volumetrics computations done separately on each block. The steps here are (1) defining the fault blocks through geologic interpretation and (2) gridding the structure within them, normally through use of digitized hand-drawn contours.

Fluid Contacts

We described fluid contacts as relatively simple surfaces, but this is not always the case. For example, the OWC might be offset. This normally occurs only at faults, so the fault polygon used to displace the structural surface will also be appropriate for the OWC. If the fault-block method is used, it is only a matter of applying the appropriate OWC grid with the corresponding structure grid.

Sometimes a series of stacked water tables exists, with a series of OWCs. If the surfaces are separated enough that each can be handled individually, use the general method as appropriately modified. If there is a complex series of OWCs, play it by ear; this is a complicated problem, and common sense must be used to simplify the analysis.

Consistent Thickness Grids

Recall that the GT grids project to negative values. The NT and following grids, which are based on them, would also contain negative values. This is desirable if only one zone (i.e., the entire reservoir) is being calculated. However, if a total must be obtained through summing two or more zones or subzones, special care is needed; the program should be instructed to treat the negative values as zero when adding the grids. Otherwise a negative value can have the effect of reducing or subtracting positive thickness.

The result of such a summation is to give a grid with the correct positive values where the fluid is present and zero values where it is absent. As discussed previously, the best maps and integrations will be obtained if the grid is allowed to project negative values several grid intersections past the zero-line. Therefore, we must modify the thickness grid, whether GT, NT, PT, or HPT, and obtain negative values if the summation gave zeros.

One way to create the new thickness grid requires that a special ability be written into the mapping program. Here the program detects zero or Null values at grid intersections near the positive nodes. The positive values and their associated trends are used to project negative values to the zero-value intersections. The resulting grid will contain the same positive values as did the original thickness grid and projected negatives at the intersections in the vicinity of the zero-line.

A second method uses grid-to-data capability (see Appendix A). First, all intersections in the grid are converted to data records. These are then edited to remove (or change to Null) all zero values. Finally, the dataset is regridded. The positive values at the grid intersections are retained without change in the resulting grid, but intersections corresponding to previous zero values will have new values assigned. Projected from trends in the positive values, these values should be negative. If the projections do not give an appropriate zero-line, the methods discussed in chapter 9 should be used, particularly if a program-defined zero-line is available.

A zero-line might be available in many instances, as consistency will be gained if we retain the gross-thickness (GT) zero-line for the other thickness grids. Figure 10.6A shows a profile through a grid, with the arrow indicating the position of the zero-line. Once the zero-line is established for the GT grid, the remaining thickness grids cannot logically have positive values outside of that line because the GT grid defines the limits of the fluid.

Suppose the net/gross ratio grid in the vicinity of the zero-line equals zero; the nodes in that region of the NT grid will equal zero, and the zero-line will have been moved inside the GT zero-line. On the other hand, if the net/gross ratio grid (Fig. 10.6B) is positive in that region, the positive GT node values multiplied by the ratios will still give positive NT nodes. In this case, the slope of the NT grid could project a zero-line slightly outside of that used with GT (Fig. 10.6C). Although still within the same grid-square, this is a clear error because GT defines the productive limit. This difference could be so small as to not be detected on the map, but could give a greater net productive area than gross area when integrated.

In order to prevent the NT, PT, or HPT zero-lines from violating the GT line, the integration programs need certain abilities. First, we must be able to obtain the GT zero-line in a form that can be used by other options, and second, the integration option must be able to read and use this zero-line. With this information, the integration program will force the projection to pass through the line, ensuring consistency.

Net/Gross Ratio

When discussing the creation of structure or fluid-contact grids, we pointed out that incomplete information, whether due to short wells or the inability to make

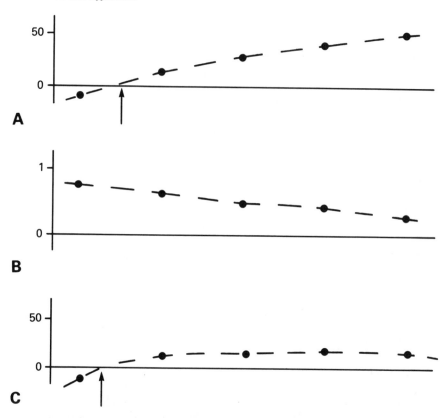

Figure 10.6 Profiles through gross thickness, net/gross ratio, and net-thickness grids. Dots represent grid-node values and dashed lines represent interpolated grids. *(A)* Gross thickness grid. Arrow locates zero-line. *(B)* Net/gross ratio. *(C)* Net thickness from product of gross thickness and net/gross ratio. Note zero-line (arrow) here has moved outside the original gross thickness zero-line.

exact picks, is a common problem. This observation led to the definition and use of limits or ranges rather than exact elevation picks. Net/gross information can be similarly incomplete.

Figure 10.7 shows various types of net/gross data. Here the central zone is of interest, and it is completely penetrated by well 2. However, wells 1, 3, and 4 appear incomplete. Well 3 is short, so the base of the zone is missing. The top of the zone in well 1 has been eroded away after deposition, while the top of the zone in well 4 has been faulted out.

The procedures for handling incomplete information can differ, depending on whether ratios or pay thicknesses are gridded. First we discuss the case of gridding ratio values, and then we look at directly gridded net thicknesses.

222

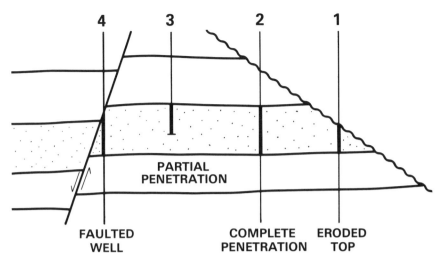

Figure 10.7 Types of net/gross data. The entire central interval is known in well 2 (thicker well bore), but is incompletely known in wells 1, 3, and 4.

Well 3 is short, so data is missing at the base of the zone. Consider the possible values that the ratio might take on if the well were complete. The missing portion of the zone might be all pay. If so, the maximum possible value of the ratio would be obtained if the missing thickness were replaced by pay thickness. On the other hand, if the missing portion were entirely nonpay, the ratio would have a minimum value if none of the missing thickness were included in the ratio. These limiting values suggest that missing data might be handled as ranges.

To put this method in mathematical terms, define the following:

$$OT = \text{observed thickness in interval}$$
$$MT = \text{missing thickness in interval}$$
$$TT = \text{total thickness in interval}$$
$$= OT + MT$$
$$NG = \text{observed net/gross ratio}$$
$$OPT = \text{observed pay thickness in interval}$$
$$= OT \times NG$$

With these definitions, the limits of the net/gross range are given by

$$\text{minimum N/G} = NG \times OT / TT$$
$$= OPT / TT$$
$$\text{maximum N/G} = (NG \times OT + MT) / TT$$
$$= (OPT + MT) / TT$$

223

These limits are then used as previously described – that is, the grid is created through an iterative process, checking for violations of the range at each step.

Figure 10.8 shows the effect of the amount of penetration. At well 1 we observe 2 feet of sand and 2 feet of shale, for an observed ratio of 0.5, with 21 feet missing. Assuming all 21 feet to be shale gives a minimum possible ratio of $2/25 = 0.08$, while assuming all 21 feet to be sand gives a maximum ratio of $(2 + 21)/25 = 0.92$. With so much missing section, the limits are wide, and this well will have little influence on the grid. On the other hand, only 1 foot is missing from well 2, so its limits are $12/25 = 0.48$ to $(12 + 1)/25 = 0.52$. This narrow range will strongly restrict the value assigned to the grid at this well, as is appropriate because the ratio is almost entirely known.

In volumetrics calculations, faults are commonly assumed to be vertical. However, the faults do not realize this, and wells can penetrate a fault plane and lose section in a zone (well 4 in Fig. 10.7). The true ratio cannot be calculated, but it can

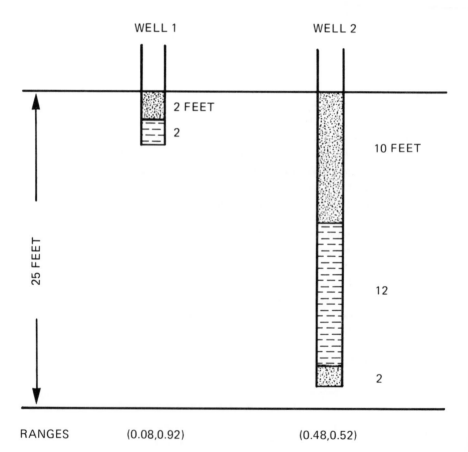

Figure 10.8 Effect of amount penetrated on width of net/gross ranges.

be converted into an acceptable range as discussed above and gridded as with partial penetrations.

At eroded regions, gross thickness can be calculated in such a way as to take truncation into account so there is no missing information, but there still can be difficulty. Figure 10.9A shows a zone that is 20 feet thick. At well 2, 2 feet of shale and 18 feet of sand give a net/gross ratio of 0.90. At well 1, the same 2-foot shale is found, but the gross thickness is 10 feet, for a ratio of 0.80.

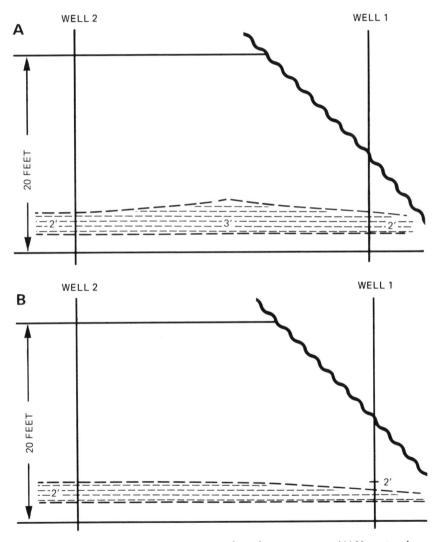

Figure 10.9 Errors in net/gross ratio resulting from truncation. (A) Honoring the observed ratios causes the shale bed to be too thick midway between wells. (B) Using the range at well 1 causes shale to thin in wedge zone.

225

If the values 0.90 and 0.80 are mapped normally, the value 0.85 will be found midway between the wells. Here the zone is 20 feet thick, so the shale is mapped to be $(1 - 0.85) \times 20 = 3$ feet in thickness, as indicated in the figure. The shale unit is thus mapped poorly with ratios, even when the observed ratios are honored at both wells, because of the rapid changes in gross thickness.

If the truncated area covers a wide region and its edge is near well 2, this effect is reduced. However, if the truncation area is narrow relative to the well spacing, the error may be significant. In such a case, it might be better not to honor the ratio at well 1. Instead, we can honor sand thickness near the well by projecting ratios from surrounding wells into the location of well 1; a range is defined at well 1 to keep the grid under some control. This range could be specified as described above (with projected pretruncated thickness giving total thickness), or percentage limits around the observed ratio could be used. With ranges, correct sand and shale thicknesses will project from well 2 to a point midway between the wells, and then sand will thin underneath the truncation, as shown in Figure 10.9B. For a limited truncation area, this procedure may give a reduced error.

If it is necessary that all observed ratios be honored, a third method is available. This method is more complicated than the two above and will be in error in the truncation (wedge) region away from the wells. The procedure is as follows: (1) Separate the wells penetrating the truncated area from those not in the truncated area; (2) grid the wells that are outside the truncated area; (3) change all grid intersections that correspond to the truncation area to Null values; (4) convert the grid to a set of data values, with the Nulls deleted; (5) combine the wells in the truncation area with the data created in Step 4; (6) grid the combined data. Again, if the truncation area is small, errors will be limited, although there may be too few wells to create a good grid.

Now let us consider the case of mapping pay (or nonpay) thickness directly. If a well does not completely penetrate a zone, we do not know whether the missing rock is pay or nonpay, but as with ratio data, ranges may be used. Again, we define the limits by assuming the entire missing thickness to be, in turn, all pay and all nonpay. If pay thickness is being mapped, the minimum thickness limit is the observed pay thickness, and the maximum thickness limit is the sum of the observed pay thickness and the missing thickness. Faulted wells with missing section may also be mapped with ranges.

As with mapping ratios, truncations can cause difficulty. Consider a zone that is uniformly 20 feet thick (Fig. 10.10A). At one point — say, well 2 — sand makes up 15 feet and shale 5 feet. At truncated well 1, erosion has left only 1 foot of sand and 1 foot of shale. If we grid sand thickness normally, it will range from 15 feet at well 2, through 8 feet midway between the wells, to 1 foot at well 1. For most geologic configurations, this interpolation is clearly in error throughout the region between the wells.

Now consider treating well 1 as a range (Fig. 10.10B). The minimum and maximum sand-thickness limits could be based on the gross thickness (which gives a very broad range in this case) or on some percentage of the observed thickness;

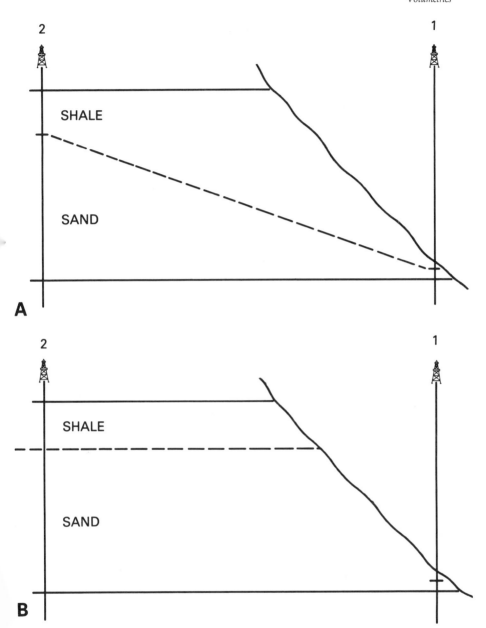

Figure 10.10 Errors in net sand thickness resulting from truncation. (A) Mapping thickness directly gives unrealistic grid between wells. (B) Using broad range at well 1 gives unrealistic sand thickness at well 1.

the range is used primarily to keep the grid from projecting unreasonably. In this case, the sand thickness from well 2 projects the value 15 toward well 1, which will probably not violate the wide range.

At some point between the wells (intersection of dashed line with unconformity), the net sand thickness will exceed the gross layer thickness. The sand thickness therefore must be altered to be consistent with the gross-thickness grid. Operating the two grids together, and retaining the minimum thickness of the corresponding intersection values, gives more realistic results. However, throughout the corrected region, the ratio would be 100%, clearly an error.

The complex third method discussed previously for ratios could also be used here; again, all data-point thicknesses will be honored, but the wedge area can be in error away from the wells. The method here is the same as described above, except that ratios are replaced everywhere with pay thickness.

Porosity

Gridding and mapping porosity leads to many of the same considerations and complications as does working with net/gross ratio data. Figure 10.7 showed various types of possibly incomplete data. Porosity information is normally missing everywhere that the ratio data is missing, so the same problems exist. In addition, porosity information can be missing from portions of the well in which net pay exists.

A common procedure for obtaining porosity is as follows: (1) Obtain measured core porosity and log values corresponding to the core depths; (2) relate the log values to the core values, commonly through a cross-plot and line fit; (3) use the relation to calculate porosity on a detailed basis; (4) use the resulting porosity estimates for mapping. However, when applying the relation, some log values may not be available (e.g., cycle skips in the sonic log), so an otherwise complete well could have several feet of information missing.

If porosity is missing, whether because of causes shown in Figure 10.7 or sampling gaps, considerations as discussed with net/gross ratio data are applicable. An immediate solution would seem to be the use of ranges. This can be a useful technique, although there are difficulties here that do not exist with net pay. When ranges on net/gross ratio are defined, two very specific choices for use in the missing interval—pay or nonpay—are available, but comparable choices do not exist for porosity.

The porosity range can be defined as follows:

$$\text{minimum porosity} = (\text{OPT} \times \text{AP} + \text{MPT} \times \text{MINPOR}) / \text{TPT}$$
$$\text{maximum porosity} = (\text{OPT} \times \text{AP} + \text{MPT} \times \text{MAXPOR}) / \text{TPT}$$

where

OPT = observed pay thickness in interval
MPT = missing pay thickness in interval

TPT = total pay thickness in interval
  = OPT + MPT
  AP = observed average porosity of pay in interval
MINPOR = minimum acceptable porosity in missing pay interval
MAXPOR = maximum acceptable porosity in missing pay interval

Application of the ranges would be the same as with net/gross ratio.

However, an immediate problem is the selection of the two porosity values, MINPOR and MAXPOR. Inasmuch as average porosity is based only on rock designated as pay, MINPOR = 0 is an unrealistic indicator of the lowest possible porosity. Simply using the lowest observed porosity value in the well will give a larger value, but the value will probably still be too low to represent an average.

Another method selects the average porosity for the interval from all wells in the dataset. This value would be better, but only if porosity trends do not exist over the area. Where there are trends, selection of the lowest average porosity from wells in the neighborhood of the one being studied is probably appropriate. This value can be obtained by averaging nearby data values, but a convenient method is to create a very coarse grid of average porosity and interpolate a value at the well location. Similar considerations apply to selection of MAXPOR.

We have found that this procedure often gives ranges with wide limits, allowing the ranged-wells to have little or no influence on the final grids. If many wells have wide ranges, it would seem unnecessary to use limits and a simpler method could be used. Several possibilities exist.

One possible method is to simply delete wells with missing information and let the gridding algorithm project values at these locations. This method is most applicable if only a few wells have missing data and have broad ranges. Going to the other extreme, we could assume that the interval is homogeneous to the extent that the missing average porosity equals the observed average porosity in the well. In this case, we would ignore the fact that some intervals are missing and create the grid with all observed data.

Another method combines interpolated and observed data. Rather than simply ignoring the observed porosity or the missing interval in the well, this method uses information in proportion to that part of the interval observed. Only the complete wells are used to create a grid, and values are interpolated at the incomplete wells. The interpolated and observed porosities are averaged, weighted by corresponding thicknesses. Thicker observed intervals thus have greater influence than thin intervals. These values are combined with the complete data, and average porosity is regridded.

How do we choose which method to use? This depends on the data, geology, and required accuracy, so no specific guidelines can be given. However, if the decision warrants extensive effort, simulations can be conducted. Using only the complete wells, do the following: (1) Remove a given proportion of the data in a well; (2) use the other complete wells to calculate the value for this well by use of several alternative methods; (3) compare these estimates to the known observed

229

value at this well; (4) repeat the process for all wells; and (5) repeat the process for different proportions of removed data.

Select the method to be used on the basis of which gives the best predictions of missing intervals. It is possible that different methods would be implied for different degrees of missing interval; if so, it would be feasible to use several methods when building a grid. However, keep in mind that it is possible to overwork the problem. Of course, a similar procedure and gridding method could be used for net/gross ratio, but there is usually little difficulty in defining a range.

Inner Wedge

In our discussion above on gross thickness, Figure 10.2B represented the horizons continuously, as they exist in nature. However, computer grids approximate these surfaces, and grids are discontinuous or discrete. Because of this nature of grids, small inaccuracies can arise when doing volumetrics at the so-called inner wedge.

Figure 10.11 shows a detailed view of the intersection of the reservoir top, T, and gas-oil contact, GOC, shown in Figure 10.2B; this intersection is at the outer wedge. Analysis of the grid intersections shows that the difference $T - GOC$ gives a continuous gradation between positive and negative values, without any abrupt changes in slope, so accurate volume calculations can be made here.

Figure 10.12A shows the inner wedge, where the GOC intersects the reservoir base, B. Here the grids are combined by lapping B onto GOC. In the resultant grid, we see an abrupt change in slope across the intersection (Fig. 10.12B). Further, the actual intersection of the surfaces does not occur at a grid node, so the change in slope is not at the actual location.

For the moment, assume that the program interpolates linearly between grid intersections. In this case, the combined grid would be handled correctly away from the true intersection but would erroneously follow the dashed line in Figure 10.12B. The combined grid therefore always projects too shallowly at this location, and integration would give a value that is slightly low. Even if a curvilinear fit rather than a straight-line segment is used, the program will underestimate volumes.

Figure 10.13 shows a cross section containing the gas column. We can create a grid of thickness above the GOC by $TH1 = T - GOC$; this is marked by the hatchured region in the figure. Integration of this grid gives the corresponding volume, V_1. Now create a grid of thickness above GOC and below B, as $TH2 = B - GOC$ (this is the double-hatchured portion), and integrate volume V_2. Note that neither of these operations involves an inner wedge configuration. Subtracting $V_1 - V_2$ leaves the desired volume of gross gas in the reservoir interval.

Now consider the more complex situation of gross oil (Fig. 10.3). Here the inner wedge is at the intersection of B and OWC, and the same configuration appears where GOC intersects T. Four steps are required here.

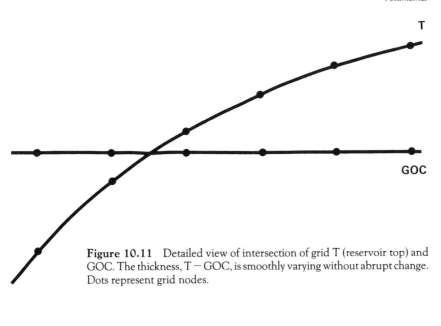

Figure 10.11 Detailed view of intersection of grid T (reservoir top) and GOC. The thickness, T − GOC, is smoothly varying without abrupt change. Dots represent grid nodes.

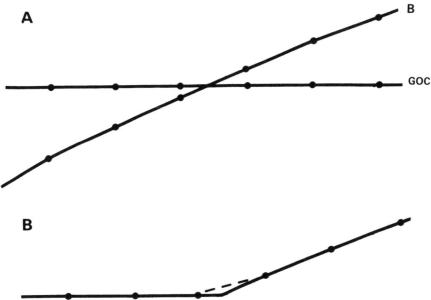

Figure 10.12 Detailed view of grids at inner wedge. (A) Intersection of grid B (reservoir base) and GOC. (B) Combined grid resulting from lapping B onto GOC. Note the abrupt change of slope at the intersection; the dashed line represents the erroneous interpolation. Dots represent grid nodes.

231

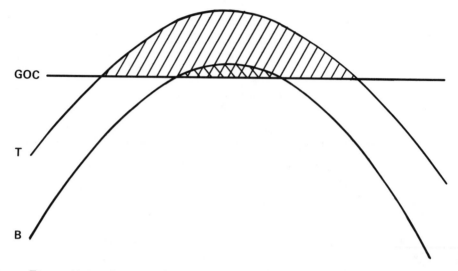

Figure 10.13 Gas interval in reservoir of Figure 10.2. Single-hatchured region indicates thickness above GOC, and double-hatchured region indicates thickness between B and GOC. Subtraction of corresponding volumes gives gross reservoir gas volume without inner-wedge problems.

1. Create grid $TH1 = T - OWC$; integrate to get V_1.
2. Create grid $TH2 = T - GOC$; integrate to get V_2.
3. Create grid $TH3 = B - OWC$; integrate to get V_3.
4. Create grid $TH4 = B - GOC$; integrate to get V_4.

Then compute the oil volume in the reservoir as $V_1 - V_2 - V_3 + V_4$. Again, no inner wedge configurations appear, but four time-consuming integrations are required.

Use of grid-equivalent methods can reduce problems with the inner wedge if finely spaced integration points are interpolated before being operated against other variables. This procedure minimizes the extent of error (dashed line in Fig. 10.12), giving more accurate integrations.

Alternative Methods

The preceding discussion summarizes a general method for detailed volumetric calculations. Because it is general, factors were discussed that may be unimportant to a given project. It is therefore possible that the process can be simplified and shortcuts used in any given study.

A classical method of calculating net volume by hand is to determine cumulated total pay thickness at each well and then to map pay thickness (cf. Allen,

1964). The planimeter is used to integrate the map to obtain volume. This method can be extended to incorporate HPT by cumulating porosity and hydrocarbon saturation with the net pay thickness. HPT is then mapped, and volumes are determined. As pointed out previously, all information must be known at each well. The procedure also ignores detailed variation in the vicinity of wedges or pay pinch-outs. However, it can give a reasonable estimate if many wells are available.

The minerals industry commonly uses a similar method. The situation is different because only a single variable (ore grade) is involved instead of three. Here the cumulative ore-grade thickness at each hole is calculated. This data could be gridded and integrated as above, but with the typical dense data control, kriging is commonly used to either develop a grid or calculate a global volume estimate (cf. David, 1977; Journel and Huijbregts, 1978; Clark, 1979).

Returning to petroleum and the general procedure, we might combine two of the variables (e.g., porosity and hydrocarbon saturation) into a single grid, or we might assume a constant value for one of the variables. The modification of the process is evident.

It is also possible to map net pay thickness directly and bypass gross thickness grids and individual ratios at wells. In Figure 10.2B, dividing the gross gas thickness grid by the reservoir isochore (T − B) grid gives a ratio grid. This ratio grid is negative where gas is absent, equals 1.0 where gross gas thickness equals the isochore, and is between zero and 1 in the wedge area. Multiplication of this grid by net pay thickness gives an estimate of net/gross ratio.

As with many of the topics in this book, several pathways to a solution exist. Some are shorter than others, and some are more accurate, but geological reasoning, interpretation, and common sense should guide the user in volumetrics.

233

CHAPTER 11

Trend Analysis

Systematic variation in rock properties over some map area is commonplace, and such spatial variation can be pictured on contour maps. However, local fluctuations may obscure underlying broad patterns, or trends may hide small anomalies. Interpretation is aided if we are able to separate local detail from the broad features.

The concepts of regional trends and local anomalies have been used for years to distinguish between components of a map representing large and small features. For example, a basin may be represented grossly as a large egg-shaped depression (regional trend) and have superimposed on it a variety of folds and faults (local anomalies). Of course, such forms may range from extremely large to very small in size, so "regional" and "local" are relative terms dependent upon map scale and the needs of the project.

Trend-surface analysis was described in general terms by Grant (1957) and applied to geology by Krumbein (1959). This concept divides observed values into regional and local components. The form of the regional component is approximated by a deterministic (usually polynomial) function in two dimensions, but several other methods are available to describe the broad features. Accordingly, the term *trend analysis* is commonly used to distinguish the general procedure from the specific polynomial trend-surface analysis, although the same approach can be used.

Trend analysis is a four-step process. The first step is to determine the appropriateness of the data and area for trend analysis. The second step is to produce a smooth map or gridded surface that describes the broad trend. Before computers,

235

this was done by drawing smooth contours on a map of the surface (e.g., Rich, 1935). The trend surface can now be generated in a variety of ways with the computer. Mathematical functions of the X and Y coordinates can be fit to the data, or moving-average or filtering techniques can be used to simplify a grid. The end result of this step should be a smoothed version of the original data in grid form.

The third step in trend analysis is to produce residuals between the smooth regional trend and the original data or map. Before computers, cross-contouring was used to produce a map of residuals by subtracting a contour map of the trend from a contour map of the original data. However, the method originally developed for trend-surface analysis with the computer involves interpolating values from the trend grid at each of the data locations. These values are subtracted from the original data values, and the resulting residuals are then gridded and contoured.

The fourth step is the evaluation of the trends and residuals. The goal of this analysis is to identify anomalies. The significance of an anomaly depends on the variable and the situation; it possibly represents such features as structural highs, avenues of fluid migration, zones of high mineralization, or faulted regions. Because of the diversity of anomalies and the regional features on which they are superimposed, several different trends are often used to produce residuals. Evaluating trend and residual maps with an understanding of the area's geologic conditions allows the most appropriate regional component to be identified. Statistical analysis aids in selecting the best trend/residual combination.

Trend analysis may be the most written-about subject in mathematical geology. Applications have ranged over all aspects of geology, geography, and geophysics. The theory and use of trend analysis are discussed by Krumbein and Graybill (1965), Merriam and Cocke (1967), Watson (1971), Agterberg (1974), Mather (1976), and Davis (1986), among others.

APPROPRIATENESS OF DATA

Before any form of trend analysis is considered, the data should be mapped and examined to determine if such analysis is appropriate and how it should be applied. For instance, does there seem to be a regional component with extreme local fluctuations that confuse the picture? Deciding how to use trend analysis depends to a large degree on the goals of the study: Is the regional trend, isolated variation, or the entire dataset of most interest?

The purpose of trend analysis is to detect, study, and possibly remove anomalies, but there may be preconceived notions or requirements as to the size of the anomalies. Anomalies must be larger than the original data spacing. Anomalies that are small relative to the data spacing might be figments of the mathematical equation used to build the trend. In any event, an anomaly defined by only a single point is questionable. Further, anomaly size can influence the choice of surface or function. A very simple surface will show large anomalies. As a surface increases in complexity, the anomalous features become smaller but more numerous.

Another consideration is whether the variable can be described by a single or simple mathematical function. For instance, a surface that has an abrupt change of slope cannot be depicted adequately by a polynomial trend equation, but a plateau may be defined as a simple flat upper surface and a simple sloping side. Each portion could be modeled separately with good results. For this reason, care should be taken to ensure that the analysis is performed only with appropriate regions and functions; it is not unusual to use subareas for separate analyses.

CREATING TREND SURFACES

Several procedures are available to detect and describe the regional trend. Common ones include fitting polynomial or Fourier series to the observed data by least-squares criteria (cf. Krumbein and Graybill, 1965; Davis, 1973). Here the parameters of the function in X and Y are adjusted in such a way as to make the function match the data as closely as possible. In addition, smoothing or filtering methods that do not fit data to functions are also available.

After the trend is created, it is necessary to convert it to a grid. If a mathematical function is used, the Z-value of each intersection is calculated by inserting the (X,Y) coordinates of that intersection into the function. For a mathematical function, the choice of grid interval will have no effect on the final surface, although the grid should be sufficiently fine that accurate inverse interpolations can be made. Smoothed or filtered grids normally have the same interval as the original grid.

Polynomial Function

A polynomial surface is a mathematical function involving powers of X and Y. Its complexity is controlled by the user through the number of terms included, which is dependent on its degree, N, a positive integer. Table 11.1 shows equations

TABLE 11.1
General Form of Several Polynomial Functions

Degree	
1	$Z = a_{00} + a_{10}X + a_{01}Y$
2	$Z = b_{00} + b_{10}X + b_{01}Y + b_{20}X^2 + b_{11}XY + b_{02}Y^2$
3	$Z = c_{00} + c_{10}X + c_{01}Y + c_{20}X^2 + c_{11}XY + c_{02}Y^2$ $+ c_{30}X^3 + c_{21}X^2Y + c_{12}XY^2 + c_{03}Y^3$
N	$Z = d_{00} + d_{10}X + d_{01}Y + \ldots + d_{0N}Y^N$
	$= \sum_{i=0}^{N} \sum_{j=0}^{N} d_{ij}X^iY^j$

for different N. A polynomial surface of degree 1 (Fig. 11.1A) is a plane. A polynomial surface of degree 2 can have only one fold in it; it can have the form of a parabola with a vertical plane of symmetry (Fig. 11.1B) or a saddle (Fig. 11.1C). Trend surfaces of degree 3 (e.g., Fig. 11.1D) and greater continue to increase in complexity.

The polynomial form is a logical choice for surface approximation as any function that is continuous and possesses all derivatives can be reproduced by an infinite power series (e.g., Taylor's or MacLaurin's series expansions). If exact details are not of interest (as they would not be for a regional trend), higher-order

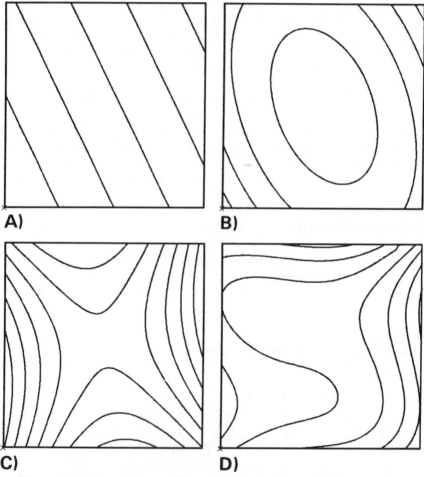

A) **B)** **C)** **D)**

Figure 11.1 Maps of polynomial functions of various degrees. (A) Plane; degree 1. (B) Paraboloid; degree 2. (C) Saddle; degree 2. (D) Degree 3.

terms can be dropped; the remaining low-order terms approximate the function.

The general form of the function is given for several degrees, N, by the equations in Table 11.1. However, the values of the constants in the equations are unknown. These constants are estimated so as to allow the function to pass as closely as possible to the data points. Most algorithms use a least-squares criterion to find those constants that minimize the sum of squares of the vertical deviations between each point and the surface. This well-known method is discussed in detail by Krumbein and Graybill (1965), Mather (1976), and Davis (1986), among many others.

For trend-surface analysis of geologic data, trends and residuals are generally computed for degrees 1 through 4 or 5. This cutoff is arbitrary, but for randomly distributed well data, degrees higher than 4 usually result in little additional information. If, however, the data are uniformly spaced or the surface varies smoothly, sixth and even higher-degree surfaces can be useful.

Table 11.2 indicates the number of terms in a degree N polynomial; the number of data points used to make the least-squares fit must exceed this number and should be several times greater if possible. The number of extrema and inflections indicate relative complexities of various degrees.

Care should be used in applying the least-squares method with polynomials. One or two points having an extremely high or low value relative to the other points can disproportionally influence the surface. Data evenly distributed over the area produce surfaces most representative of the actual trends, but clustered data having very few outlying points can produce trends that mimic the shape of the data cluster (Doveton and Parsley, 1970; Shaw, 1977). A few outlying pseudo-points added to the data can remove this effect, but these points can strongly affect the surface. The effect of data distribution is discussed by Miesch and Connor (1968) and Mather (1976).

Because the trend equation involves geographic coordinate values raised to

TABLE 11.2
General Properties of Polynomial Functions

Degree	Number of terms	Maximum number of extrema	Maximum number of inflections on profile
0	1	0	0
1	3	0	0
2	6	1	0
3	10	4	1
4	15	9	2
5	21	16	3
N	$(N + 1)(N + 2)/2$	$(N - 1)^2$	$N - 2$

powers, problems with rounding error can result. For instance, a function of degree 5 requires calculation and summation of powers of 5 for X and Y. If these coordinate values are large (typically 6 or 7 digits), severe computational problems can result due to an insufficient number of significant digits in the computer. It is best to remove the leftmost digits in the coordinates by subtracting appropriate constants from all X and Y values before applying the trend analysis program, particularly if higher-order terms are to be computed.

The calculations involved in fitting trend-surface analysis polynomials are extensive and require solving large sets of simultaneous equations, which typically have poor computational properties. In addition to these difficulties, the resultant polynomial terms are not independent, so it is not possible to separate the total surface fit into separate term-by-term components. Use of orthogonal polynomials solves these problems, as each coefficient can be appraised independently. Early use of orthogonal polynomials required that data be sampled on a grid, but Whitten (1970, 1972) discusses the application for irregularly spaced data.

A number of programs have been written to perform the calculations for polynomial trend-surface analysis. Esler et al. (1968) and Davis (1973) provide listings. Least-squares trend surfaces are special cases of multiple linear regression, so regression programs are also of interest (cf. Mather, 1976). Campbell (1974) and Mather (1976) provide programs to analyze the significance of terms of higher degree, and Whitten (1974) provides a program for calculation and use of orthogonal polynomials. Most mapping packages include polynomial trend-surface analysis.

Double Fourier Series

The double Fourier series consists of harmonic wave forms, defined by sines and cosines of decreasing wavelength. When defined as a function of (X,Y) coordinates, fundamental wavelengths (L and W) are specified in the directions of the coordinate axes. L and W ideally should be based on fundamental properties of the surface, but commonly are set equal to the length and width of the area or grid. The first harmonic has wavelengths equal to those fundamentals, and all other wavelengths are integral fractions of these: L/2 and W/2, L/3 and W/3, and so on. As with the polynomial function, an infinite series of these terms exactly reproduces a smooth function.

Table 11.3 presents the form of the double Fourier series for harmonic order n. The unknown coefficients or amplitudes are estimated from the data by least-squares, as with the polynomial function. Table 11.4 summarizes some of the properties of the function for various harmonics. As with the polynomial, the number of points used to fit the chosen function must exceed the number of terms.

In practice, a limited number of terms is adequate to portray the function, although spurious harmonic forms may show up in both trend and residual maps if too low an order is used or if the variable does not have a harmonic content.

240

TABLE 11.3
General Form of the Double Fourier Series of Order n

$$Z = \sum_{i=0}^{n} \sum_{j=0}^{n} (a_{ij}A_iC_j + b_{ij}A_iD_j + c_{ij}B_iC_j + d_{ij}B_iD_j)$$

where a, b, c, and d are unknown coefficients to be estimated and

$A_i = \cos(2i\pi X/W)$
$B_i = \sin(2i\pi X/W)$
$C_j = \cos(2j\pi Y/L)$
$D_j = \sin(2j\pi Y/L)$

and W and L are fundamental wavelengths in the X and Y directions. Note that $B_0 = D_0 = 0$.

TABLE 11.4
General Properties of Double Fourier Series

Order	Number of terms	Maximum number of extrema	Maximum number of inflections on profile	Minimum wavelength W	L
0	1	0	0	∞	∞
1	9	4	2	W	L
2	25	16	4	W/2	L/2
3	49	36	6	W/3	L/3
n	$(2n+1)^2$	$(2n)^2$	$2n$	W/n	L/n

Because the harmonics are repeating sines and cosines, the function repeats for distances greater than L and W, and the Fourier series is not appropriate for extrapolation. For a dipping surface, maps are often improved by subtracting a plane from the data before fitting the double Fourier.

The use of the double Fourier series in trend analysis was discussed by Bhattacharyya (1965), Harbaugh and Preston (1965), and James (1966), and more recently by Davis (1973), Agterberg (1974), and Robinson (1982). Applications include modeling folds (e.g., Whitten and Beckman, 1969) and cyclic paleotopography (Maslyn and Phillips, 1984). Programs have been provided by Preston and Harbaugh (1965) and James (1966). Fitting this model is similar to fitting the polynomial, so the considerations discussed by Mather (1976) are appropriate.

Moving-Average and Smoothing Methods

Polynomial functions and Fourier series are restricted to particular forms. For a given degree or order, the function can have only a specific number of highs, lows,

241

or inflections (Tables 11.2 and 11.4). These limited functions cannot reproduce the more complex shapes found in earth-science applications. Further, they are usually not appropriate for situations in which the anomalies are unevenly scattered.

An alternative method for finding a trend is the moving average. Starting with an initial grid that is created normally, smoothing operations "bulldoze" highs, "fill in" lows, and smooth small features. This process therefore reflects the general features of the mapped data, rather than an arbitrary mathematical shape. We have generally found this type of surface to be more useful in defining a trend than mathematical functions.

The moving-average method operates over the grid intersections one by one. The user selects a window, commonly a circle with specified radius. Figure 11.2 shows a circle of radius 2 grid intervals. This circle is centered over a given grid intersection. All 13 intersections within the window are then averaged, and the result becomes the Z-value in a new grid for that central intersection. The window then moves to another intersection, and the process is repeated. Moving the window and averaging intersection Z-values gives the process its name.

Rather than a simple average of the intersection values, a weighted average is commonly calculated. In this case, weights are assigned to each of the intersections within the window circle, and these weights are combined with the values when calculating the average. The usual procedure gives higher weights to intersections near the center of the circle and lower weights around the periphery. The weights can be selected to define a filter with specified properties.

The user must specify the window radius. Windows with larger radii remove larger features. A minor amount of smoothing (small radius) may be appropriate for removing small bumps and wiggles from grids and contours. However, trend analysis generally requires a great deal of smoothing (large radius). If a program has a radius limit that is too small, a grid that has already been smoothed should be

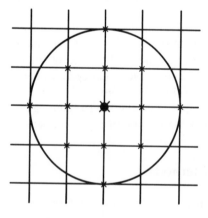

Figure 11.2 Moving-average smoothing with radius of two grid intervals. All 13 grid values within the circle are averaged, and the result is assigned to the corresponding central intersection in a new grid. The circle is moved in turn to every grid intersection, and the process is repeated to create a new grid.

smoothed again. This two-step process affects intersections at a distance corresponding to the sum of the radii.

Filtering

As with smoothing, filtering is used to define trends and residual grids with properties that are too complex for simple functions. In addition to obtaining anomalies of specified size, shape, or orientation, the use of digital spatial filters allows us to obtain a trend grid with unwanted features removed. Besides giving an uncluttered look at the regional trend, it produces residuals with specified characteristics.

Digital filters come from communications theory and are based on harmonic and spectral analysis of waveforms. An observed signal is assumed to consist of a mixture of sinusoidal waves of different frequencies. A filter is a mathematical operation that will remove specified wavelengths. For example, the filter may be designed in such a way that high frequencies (noise) are removed and low-frequency components (signal) are passed, for a specified cutoff frequency. Band-pass filters remove both high and low frequencies and retain intermediate wavelengths. Theoretical aspects of filters are reviewed by Lee (1969), Clement (1973), and Mesko (1984).

Application of filters in geology assumes that the geological surface or variable consists of a mixture of features with various wavelengths or frequencies, leading to a combination of signal and noise. This assumption is similar to the idea that a trend can be defined by a double Fourier series, with residuals (noise or high frequencies) superimposed. Filters can thus screen out the noise, leaving the trend. Earth-science applications of filters are discussed by Robinson et al. (1969), Robinson (1969, 1982), Spector and Grant (1970), Robinson and Merriam (1971, 1972, 1984), and Wilson (1975), among others.

A spatial filter must be applied to an existing grid, as with smoothing. The filter is an array of values, much like a small grid with the same interval as the grid being analyzed. This array should be much smaller than the grid, as the search circle described above is small relative to the extent of the grid. The size of the array and the values used in it are determined by the required properties of the filter.

There is a great deal of flexibility in the design of the filter. It can be defined in either the space domain (in which case it modifies the surface on the basis of size) or the frequency domain (in which case it operates on frequencies). A single size or frequency cutoff, or multiple cutoffs, may be used. This can be done isotropically, or we might specify that specific directions are important. This will increase the importance of features that are so oriented and decrease or remove features with a different orientation.

Figure 11.3 shows an example of (A) a simple north-south oriented filter and (B)

243

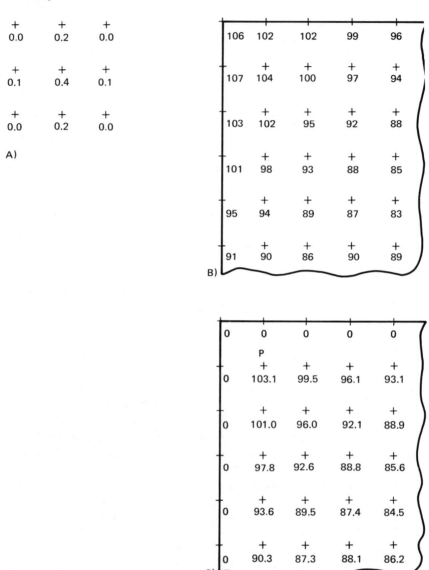

Figure 11.3 (A) North-south-oriented digital filter. (B) Portion of an isopach grid. (C) New grid that results from application of filter to isopach grid. Value at point P is $Z(P) = 102 \times 0.2 + 107 \times 0.1 + \ldots + 100 \times 0.1 = 103.1$

a portion of an isochore grid. The filter is applied to each intersection in the grid one by one, as with the moving average. For a given grid intersection P, the filter values and the local grid values are combined to calculate a new value. This is done as a sum of the grid Z-values, weighted by the filter values W. At point P we calculate

$$Z(P) = \Sigma \ W_i Z_i = 103.1$$

This value is then placed in the new grid at the intersection corresponding to P. The same procedure is followed for the next intersection, and so on until the new grid is built (Fig. 11.3C).

Note that the new grid has a row and column of zero values along its edge. Study of the process shows that it is not possible to perform the operation at the edge of the grid, so application of the filter creates an edge effect equal to the radius of the filter (here, one grid interval). For this reason, filters should be made small relative to the grid, although this small size impacts the filter's performance.

Applying a filter to a grid is a simple operation. However, designing the filter to have desired properties is more complex. Defining a digital spatial filter consists mainly of harmonic and spectral analysis (Agterberg, 1974), which determines the frequencies that are present and those that are to be filtered. Creation of the filter largely involves selection of the features to be enhanced (Fuller, 1967).

Geologic filters used for trends have some restrictions. In order to retain amplitudes, the weights must sum to 1. Also, because we do not wish to displace features laterally, the filter should be zero-phase — that is, an even function. (In terms of waveforms, it consists entirely of cosines.) The filter thus has an odd number of rows and columns. The filter with radial symmetry operates isotropically, retaining features regardless of orientation. Filters with axial symmetry (e.g., Fig. 11.3A) take into account the orientation of the features. Wilson (1975) discusses these fan-pass filters.

Robinson (1982) points out that many geological studies can be conducted with only a few standard filters: regional, directional, and isotropic. Adjustments of these can be made through scaling and then applied to many situations. Robinson et al. (1969) and Robinson (1982) discuss the interpretation of filtered maps.

Nonlinear Functions

The polynomial and double Fourier series functions do not attempt any geological significance; they are only meant to provide a good representation of the data. The constants associated with the terms in the functions do not have any physical significance, but simply define the surface (although the constants associated with a plane fit might be used to define dip and strike). Similarly, moving-average methods operate on grids without regard to an underlying model.

245

In some situations a model with specific form might be more appropriate. For instance, James (1968) studied beach cusps and found that this complex form could not be well represented by either polynomial or Fourier series. Further, the wavelength and amplitude of the cusps were important for analysis, but they are not described by terms in those series. He therefore developed a geometric model containing parameters that have physical meaning.

Some models with physical meaning will be linear in the coefficients; an example of this is the incorporation of a simple fault into a polynomial surface (James, 1970; Attoh and Whitten, 1979). However, more commonly a realistic model will be nonlinear. James's analyses of "low and ball structure" in nearshore profiles (1967) and beach cusps (1968) are examples of nonlinear modeling. Sloss and Scherer (1975) simulated a sedimentary basin with an inverted bivariate normal curve.

Nonlinear models are not handled by standard trend-analysis programs, and least-squares fits must be programmed individually. James (1968) presents an iterative program, and other methods are summarized by Mather (1976). In most cases, the goal of such a fit is to obtain a good set of parameters, so residuals are usually analyzed in order to determine if the fit is adequate. However, further analysis of residuals can also be made.

CREATING A RESIDUAL GRID

Regardless of whether the trend grid was created by a mathematical function, filtering, or some other procedure, the next step is construction of a grid of residuals. Two choices are available, depending on whether the trend was calculated from the original data (fit function) or from a grid (smoothing or filtering). The following steps outline how residuals are calculated when the trend is created from the original data.

1. Interpolate a value from the trend grid at each data location. Combine these values with the original data.
2. Calculate residuals by subtracting the value representing the trend from the corresponding original data value.
3. Generate a grid of the residuals calculated above.

If the trend grid was created from another grid, residuals can be calculated by the following steps.

1. Calculate a grid of residuals by subtracting the trend grid from that based on normal gridding methods.
2. To determine the residual value at each data point, interpolate a value from the residual grid at each data location.

This method is used if smoothing or filtering was done, as they operate on an initial grid. It is possible to fit a trend surface to node values in a grid. In this case, this method would also be appropriate.

Filters can also be used to create a residual grid directly, again with specified orientations or dimensions of the retained or filtered features. When defining the filter, the sum of the weights must equal zero. This residual grid can be analyzed itself, or it can be combined with the original grid to define a trend.

EVALUATION OF THE TRENDS AND RESIDUALS

One purpose of trend analysis is to detect and interpret areas of deviation (i.e., residuals) from the "regional" or "trend" component in the data. Maps of these residuals show areas of negative and positive contours that define anomalies. It is the task of the geologist to interpret these residuals and determine if they are significant. For instance, residuals that are limited areally might be too small to be important. Similarly, large regions of residuals on the map, but with amplitude values near zero, might also be unimportant. These results can also indicate that the trend is too complex, and that a simpler, smoother trend should be used.

If we generated several degrees of trend surface, or used several different methods, we must select a single trend. Selection is normally based on geologic considerations, but statistical analysis can help. Stepwise regression procedures point out which additional terms in polynomial or double Fourier series fits are significant (e.g., Graybill, 1961; Draper and Smith, 1966; Mather, 1976). However, be cautious, as certain statistical assumptions will not be met unless the entire trend has been removed and the residuals are random.

Selecting the trend-residual combination to use is normally not too difficult, as a particular trend will usually stand out from the others. However, if the choice is not obvious, or if a better understanding of the trend surface is needed, statistical evaluation of the residuals can be made.

Several statistics can be calculated on the set of residuals for each trend. A common choice is the standard deviation or variance of the residuals. As a consequence of least-squares, the variance must decrease in value for greater numbers of terms in the trend, whether polynomial or Fourier series. Similarly, the sum or mean of the residuals must equal zero. If this is not the case, suspect rounding errors in the computer or difficulties in interpolating from the grid. Significant decreases in variance as N is increased indicate that the added terms are probably necessary. A small change indicates no need for higher degrees.

A statistic commonly used for this type of work is R^2, or the coefficient of determination. R^2 calculates the percent of the variability in the mapped variable that is accounted for by the trend. Larger values indicate better degrees of fit of the surface to the data. As more polynomial or Fourier terms are included, R^2 must increase toward 100%. Again, small increases in this statistic indicate no need for higher-degree trends. (A decrease in the statistic for larger N indicates a problem, as with an increasing variance.)

Calculate R^2 by

$$100 \, (\text{Var}(Z) - \text{Var}(R)) \, / \, \text{Var}(Z),$$

where $\text{Var}(Z)$ is the variance of the original data Z-values and $\text{Var}(R)$ is the variance of the residuals. Similar calculations can also be made for trend grids from smoothing or other methods, but statistical properties of nonfit methods will differ; for instance, the residuals will not sum to zero. Mendelbaum (1963) suggests that a better separation of trend from residual may be given by the minimum mean-square deviation.

Other useful information may include the size of the extreme residuals. A residual value exceeding a certain level may be unrealistic. If any (or too many) large residuals exist, a different-order trend should be considered. Large residuals can result from a too-simple trend function that does not represent an extreme data value or from a too-complex surface that goes out of control. On the other hand, if the extreme residuals are near zero, this trend may lack anomalies, suggesting that a simpler surface would be appropriate.

EXAMPLE APPLICATION

To demonstrate trend-surface procedures, we use the structural elevations of the unconformity surface (UNCF) mapped in Figure 4.3. This surface is moderately irregular and is observed at the evenly spaced control points listed in Table B.4. Polynomial trend surfaces for degrees 1 through 8 were constructed as described above. Figure 11.4A-C shows resulting maps for degrees 3, 4, and 5, each with a contour interval of 50 feet. The surfaces increase in complexity (number of extrema and inflections) with higher degree. The fifth-degree trend also shows slight unrealistic edge effects. A map created by smoothing (radius = 8 grid intervals) is shown in Figure 11.4D, which is contoured with an interval of 50 feet.

The residuals at the data points were calculated by interpolating values from each trend grid and then subtracting these values from the observed elevations. The residuals were then gridded and contoured normally. Figure 11.5 shows residual maps for the trends in Figure 11.4—that is, degrees 3 through 5 and the smoothed grid. All these are contoured at an interval of 20 feet.

Table 11.5 shows statistics on the calculated residuals. Goodness of fit, measured by R^2, increases with increasing degree, the rate slowing for degrees above 5. This implies that most variation in the data is honored by terms of degree 5 and lower; indeed, 80% of the variation is accounted for by degree 3. R^2 thus indicates that degree 3 accounts for much variability, but that adding fourth- or fifth-degree terms might improve the fit locally. The standard deviations of the residuals and

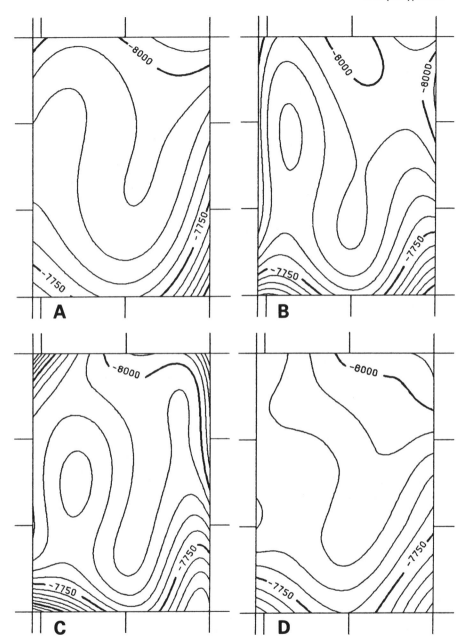

Figure 11.4 Trend maps, with contour interval of 50 feet. Data are the same as in Figure 4.3. (A) Third-degree trend. (B) Fourth-degree trend. (C) Fifth-degree trend. (D) Trend made by smoothing with a radius of eight grid intervals.

249

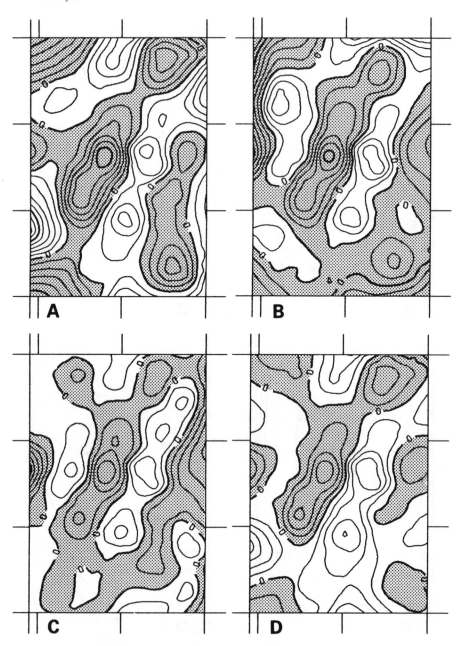

Figure 11.5 Maps of residuals from trends in Figure 11.4. (A) Third degree. (B) Fourth degree. (C) Fifth degree. (D) Residual from smoothed grid.

TABLE 11.5

Statistics on Structural Elevations of Unconformity and Residuals from Trends

	Mean	Variance	Standard deviation	Minimum value	Maximum value	R^2
Original data	−7897	8924.1	94.47	−8059	−7567	−
Residual from						
Degree 1	−0.021	5616.2	74.94	−146.9	246.4	37.1
Degree 2	−0.037	2896.2	53.82	−110.0	152.6	67.5
Degree 3	−0.050	1779.9	42.19	−107.5	114.2	80.1
Degree 4	0.013	1049.7	32.40	−66.1	105.7	88.3
Degree 5	0.062	608.5	24.67	−53.3	75.7	93.2
Degree 6	−0.108	454.4	21.32	−47.4	69.2	94.9
Degree 7	−0.187	232.6	15.25	−39.3	41.2	97.4
Degree 8	0.093	84.4	9.18	−18.6	23.6	99.1
Smoothing, radius 10	−15.520	1543.1	39.28	−97.6	80.1	82.7
Smoothing, radius 8	−10.484	1052.3	32.44	−79.5	73.0	88.2
Smoothing, radius 5	−4.283	373.8	19.33	−49.7	50.9	95.8

the minimum and maximum residuals lead to similar conclusions. A trend of degree 3, 4, or 5 would seem most likely to produce meaningful results.

Study of the mean residuals in Table 11.5 shows values acceptably near zero, as required by least-squares procedures. However, degrees 6, 7, and 8 show greater deviations, probably because of machine rounding errors or difficulty with inverse interpolation in areas of extrapolated steep slopes. The mean residuals from smoothed grids need not be zero, so the R^2 values are not directly comparable to the polynomial fits, but they can be used as guidelines.

Statistical evaluations of trends and their residuals do not have direct geological implications, although they serve as useful guides to selecting the degree of a surface. However, statistical tests are no more than guides, as certain assumptions in trend analysis violate assumptions in regression analysis. The best trend is the one that most accurately represents the geologic situation, regardless of statistics.

With this in mind, compare the residual maps in Figure 11.5. Compared to the others, the residuals from smoothing (Fig. 11.5D) are small, indicating that small features in the data have been removed; this trend is probably of no value for detecting anomalies. Compared to the other two polynomial surfaces, the fifth-degree trend has removed small-scale variation. Residuals here are of lower amplitude and are smaller areally and scattered irregularly; interest in small residuals would point us toward the fifth-degree trend.

251

There is not a great deal to choose between the third- and fourth-degree trends. The third-degree residual map is composed mainly of anomalous regions, whereas the fourth-degree residuals are not quite as abundant, although both sets of residuals are similar in form and nature. However, the third-degree trend map seems to represent the original information (and Fig. 4.3) better. While these sets of maps suggest some conclusions, additional geologic analysis should be brought to bear.

OTHER USES OF FILTERS AND FILTER ANALOGUES

In addition to defining trend surfaces, filters or their analogues can be useful for detection or enhancement of trends. Even if residual analysis is not of interest, filtering a surface can improve its regional character. For instance, data distribution and original mapping might have led to an erratic and noisy map, as with data from petroleum seismic surveys. An extremely erratic map hides general features that may be of interest.

A filter can be used to remove the small features, leaving a clearer regional trend. The filter would be designed with a cutoff size such that features of economic interest would be retained. Similarly, we might wish to accentuate or diminish features that are oriented in a particular direction. For instance, the regional structural framework might imply that en echelon folding with a specific orientation is to be expected, so we would design the filter to stress such features.

We may wish to enhance specific features but remove both regional and small residual effects. For instance, Robinson (1982) and Robinson and Merriam (1972, 1984) discuss filters that leave only features with a specified median width. Application of such a band-pass filter allows the potential structures to be detected; further interpretation with the original structure map would then be appropriate. The filter must be specially designed for the width of interest, of course, and any desired directional properties.

Another purpose of filters involves smoothing a grid. However, such moving-average methods primarily plane highs off the surface and raise the lows, but irregularities at intermediate values may not be affected. Such smoothing is useful for defining a trend but would not meet a goal of removing small-scale irregularities from a map.

Briggs (1974) and Swain (1976) discuss a procedure whereby a grid can be modified in such a way as to minimize local curvature. The process was defined mathematically as fitting two-dimensional spline functions, but it can be put into the form of a filter operation. The operation must be applied iteratively, but it works effectively to remove small bumps and wiggles from the contours and does not affect the highs and lows as drastically as moving averages. This procedure is routinely used after grid generation.

Slope mapping is another application of filtering. Usual contour presentations emphasize highs and lows and the patterns that can be built from such features.

Slope mapping, however, stresses steep gradient zones. This is useful for mapping and interpreting gravity or magnetic-intensity data on a broad regional basis. Boundaries between large blocks or regions lead to high gradient areas, which are indicated by clusters of contours. Slope mapping is a way to enhance and detect subtle changes in gradient.

Dole and Jordan (1978) create slope maps with filter operations. Their process is to (1) calculate the degree of local slope at a given grid node through analysis of nearby node values; (2) assign the value to the corresponding node; and (3) repeat for all nodes, thereby creating a new grid. The calculated value is the magnitude representing the dip in the direction of steepest gradient, calculated by combining the slopes in the X and Y directions. Because the filter emphasizes high frequencies (Zurflueh, 1967), noise can be influential, but useful grids result from this filter.

Robinson (1982) discusses other applications of filtering that can enhance specific aspects of trends in data. These include first- and second-derivative and gradient maps that detect features similarly to slope maps. Bornemann and Doveton (1983) used a second-derivative map of magnetic-field intensity to detect lineaments. Gravity and magnetic-field data commonly require upward or downward continuation to give a different datum, and filters can be designed for this. Filters can also be applied to previously filtered grids.

CHAPTER 12

Historical
Reconstruction

Chapter 4 covered the creation of grids and contour maps, and chapter 6 the extension from single surfaces to complex stratigraphic frameworks. There we saw that stratigraphic and structural relationships must be taken into account if the geology of an area is to be modeled correctly. Such concepts as truncation, baselap, and conformability allow creation of sets of internally consistent grids and corresponding contour maps. These grids represent present-day geology.

In addition to mapping the presently observed structural forms of geologic surfaces, modern quantitative procedures make it increasingly important to reconstruct detailed geologic history. Various periods of diastrophism, deposition, onlap, downlap, nondeposition, erosion, and faulting strongly influence the present configuration. Full understanding of the present geology requires knowledge of the complete history, so all of these influences must be considered.

For petroleum applications, time of hydrocarbon migration, pathways from source to trap, time of trapping, and hydrocarbon quality and quantity are more predictable by an intelligent historical interpretation. Modeling pre-erosional and posterosional configurations and thicknesses may provide insights into the timing, formation, and destruction of traps. Such analysis can also provide explanations of why a structure is empty, as different migration pathways could have put the petroleum in another location, or leakage from the trap could have occurred after migration (cf. Paynter, 1970). The effects of erosion can similarly be studied in connection with orogeny. In coal or stratigraphic mineral deposits (e.g., sedimentary uranium), similar analysis of erosion and subsequent deposition can be important for detecting channel fills and computing volumes.

Three basic steps are used to show stratigraphic and structural relationships through geologic time: (1) Interpret major geologic events and stratigraphic-structural relationships; (2) capture the data; and (3) construct the grids and maps. These steps, discussed in the first three sections of this chapter, use many concepts and procedures discussed previously and synthesize material presented elsewhere in the book.

After the grids are made, we can use them to construct displays and make calculations, including paleostructure maps referenced to some datum, structure maps of horizons reconstructed to pre-erosional form, maps of pre-erosional thickness, paleo-cross sections, perspectives and other displays showing paleostructure or paleothickness, and burial history curves. All these are discussed in the fourth section.

STRATIGRAPHIC AND STRUCTURAL CONSIDERATIONS

A complete geologic study must incorporate all significant geologic events, and the first step in this important task is interpretation, as discussed in chapter 6. The information for interpretation comes from observation and geologic inference. In a frontier area, seismic stratigraphy (e.g., Vail et al., 1977) combines readily with mapping. In developed areas, well logs and cores provide data. But regardless of the source of information, interpretation must include correlation of significant depositional and erosional markers, determination of their relative ages (i.e., the order of occurrence of the geologic events), description of their geometric relationships, and interpretation of faulting.

The relative geometry of the various stratigraphic markers must be identified. Figure 12.1 shows a typical stratigraphic framework, with the following history: (1) Horizons A, B, and C were deposited conformably; (2) surface D is an unconformity that truncates horizons below it; (3) conformable horizons E and F lap onto unconformity surface D; (4) surface G truncates horizons below it. We must model all significant geologic events to ensure that they are correctly included in the reconstruction.

Truncation is an important consideration, and we must determine which horizons were truncated or removed by orogenic events and which horizons had so little truncation that erosion can be ignored. Also, for the purpose of reconstruction, the pre-erosional thicknesses of the intervals must be estimated.

DATA CAPTURE

The second major task is to capture data for use by the computer. The following must be entered into the computer for each data location: (1) the present elevation of each horizon and each truncating surface (unconformity) in the study; and (2) an

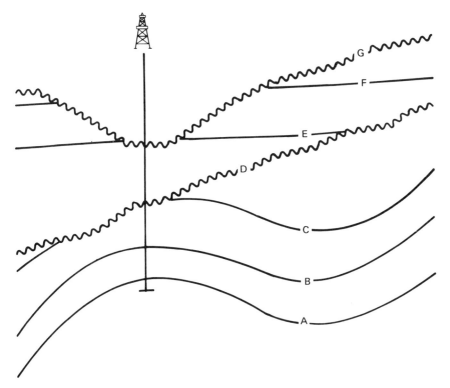

Figure 12.1 A typical stratigraphic framework, showing conformity, baselap, and truncation. (After Jones and Johnson, 1983, p. 1416)

estimate of the original thickness of each eroded interval, or of the uneroded structural elevation. Interpretation of faults is also necessary; the type of information required depends on the method used (chap. 8).

Present-Day Stratigraphy

Horizon picks are discussed in detail in chapters 3 and 5, but reiteration of two points may be useful. First, each horizon should have an observed data value except where the horizon is missing by nondeposition or erosion. In other words, the tops should represent control points on a stratigraphic surface, rather than merely the topmost rocks of an interval (unconformity picks). The well in Figure 12.1 should have tops recorded for horizons A, B, D, and G, but not for missing horizons C, E, or F. Errors result if unconformity picks are mixed with stratigraphic picks (cf. chap. 5; Iglehart, 1970). Second, fields corresponding to missing horizons should be assigned a Null value and not used in calculations.

257

Estimates of Erosion

Additional information is needed to reconstruct the pre-erosional configuration of eroded horizons. This is usually the most difficult interpretation step in the procedure, particularly if control points are sparse, although seismic sections aid interpretation. The erosional information at each (X,Y) location may be supplied as reconstructed thicknesses (Method 1) or as reconstructed elevations (Method 2).

Thicknesses can be recorded in either of two ways: (Method 1A: Thickness-to-Reference) Estimate the pre-erosional thickness between the horizon of interest and a non-truncated reference horizon; or (Method 1B: Interval Thickness) estimate the pre-erosional thickness of each eroded interval. One additional data field must be added for each eroded horizon or interval if Thickness Methods 1A or 1B are used. Method 2 (Estimated Elevation) requires estimating the pre-eroded elevation of the missing horizon directly. For the estimated-elevation method, the estimated elevations can be put in the same field as the observed picks for that horizon, replacing the Null values.

Method 1A: Thickness to Reference

Figure 12.2 shows a section with extensive erosion. If Method 1A is used, estimate the thickness of rock that existed before erosion between a reference

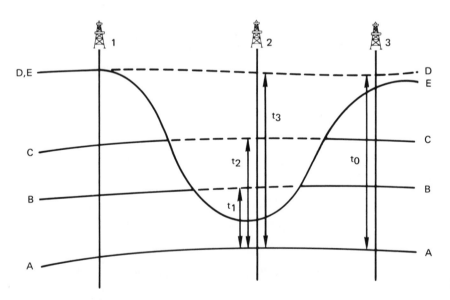

Figure 12.2 Information needed to reconstruct pre-erosional surfaces. Extensive erosion of B, C, and D requires estimation of pre-eroded thickness *t*. (After Jones and Johnson, 1983, p. 1417)

258

horizon and the interval of interest. Horizon A is chosen as reference in Figure 12.2, so we estimate the thickness between A and each of horizons B, C, and D that existed before erosion created horizon E. If there was no erosion at a given well, record the present-day observed thickness for the interval at that well. For brevity, the notation (A,B) is used to indicate the thickness from horizon A to horizon B.

In addition to the five horizon elevations at each well, three thicknesses are needed in this example. At well 3 these are the observed thicknesses (A,B) and (A,C) and estimated thickness $t_0 = $ (A,D), which is greater than the present thickness (A,E). At well 2, estimate all three previous thicknesses with the values t_1, t_2, and t_3. No erosion has occurred at well 1, so the three thicknesses are recorded as observed.

Two complications can arise in interpreting these thicknesses. The first is that the reference horizon may have been eroded at some locations, as shown in Figure 12.3. The procedure described above can be used for well 1. However, horizon A is missing at well 2, so only the elevation of unconformity surface E is recorded. The fields for A, B, C, and thicknesses (A,B), (A,C), and (A,D) are all assigned a Null

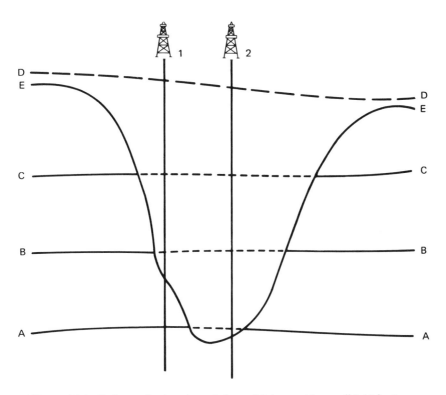

Figure 12.3 Reference horizon A eroded at well 2, but usable at well 1. (After Jones and Johnson, 1983, p. 1418)

value. If the grids extrapolate wildly in this region, an elevation for A may be interpreted and included (it will later be truncated out) and the three thicknesses estimated, although we do not recommend this procedure for most situations.

The second complication involves multiple periods of erosion, as in Figure 12.4. This history can be summarized as follows: (1) Horizons A, B, and C were deposited conformably; (2) unconformity D removed C and portions of B; (3) conformable horizons E and F lap onto horizon D; (4) unconformity G removed all or part of horizons B-F. Two erosional surfaces, D and G, are present.

For this area, thicknesses (A,B), (A,C), (A,D), (A,E), and (A,F) should be recorded. At some locations in Figure 12.4, horizon E is at a lower elevation than the pre-erosional horizon C, making thickness (A,E) less than (A,C), even though E is younger than C. For multiple periods of erosion, it is generally simpler to use only a single reference horizon (here A). However, it is possible to use a different reference for reconstructing shallower parts of the section.

Method 1B: Interval Thickness

For this method, interpret the pre-eroded thicknesses of the intervals. Returning to Figure 12.2, thicknesses (A,B), (B,C), and (C,D) are estimated and recorded. This involves the same considerations as in the thickness-to-reference method. Whichever of these two methods is used depends on the preference of the geologist and the type of information available. Again, both the observed interval thicknesses and the estimated thicknesses will be used.

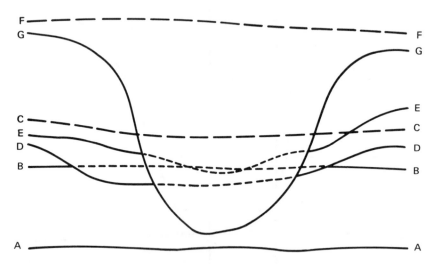

Figure 12.4 Complexities in reconstruction of B, C, D, E, and F caused by two periods of erosion (unconformities D and G). (After Jones and Johnson, 1983, p. 1418)

For multiple periods of erosion (Fig. 12.4), estimate and record thicknesses for intervals that occur in the same sequence. The first sequence in the figure consists of horizons A-C, so thicknesses (A,B) and (B,C) are used. Horizons E-F define the second sequence, so (E,F) is recorded. Missing tops in horizon D could be estimated as in the estimated elevation method, or as a thickness if D conforms in shape reasonably well to A, B, or C.

Method 2: Estimated Elevation

Here the elevations of the horizons are estimated as if they had not been eroded, referenced to the same datum (usually sea level) as the observed picks. In Figure 12.2, estimate picks in well 2 for horizons B, C, and D, and for horizon D in well 3. These picks may be recorded in the same fields as the observed picks for the corresponding horizons, replacing Null values.

CONSTRUCTION OF THE GRIDS

General Procedures

A complete historical reconstruction can result in many grids for present and past configurations. The entire task of building and reconstructing the set of horizons may appear formidable because a complex history requires a complex set of grids. However, the job can be divided into a number of simple steps if done to simulate geologic development. Keep in mind that a stratigraphic framework exists at any stage of geologic history. At a selected geologic time, no horizons younger than that time exist, so only horizons of this age and older must be considered.

The task now consists of building a series of simple stratigraphic frameworks, as discussed in chapter 6. Consider the case shown in Figure 12.2, with the corresponding framework table (Table 12.1). Before gridding is done, the information on restored erosion must be put into a suitable form. If the thickness-to-reference method is used, the thicknesses are combined with the elevation of the reference horizon to compute the estimated elevations of the eroded surface. For instance, in Figure 12.2, the following calculations would be made:

elevation(B) = elevation(A) + (A,B)
elevation(C) = elevation(A) + (A,C)
elevation(D) = elevation(A) + (A,D)

These elevations replace the Null entries in the appropriate elevation fields for horizons B, C, and D.

261

TABLE 12.1
Framework Table for Figures 12.2 and 12.3.

Surface name	Surface type	Sequence number	Sequence control	Initial grid-building method	Primary baselap/ truncate	Special baselap/ truncate
E	UNCF	X	X	DIRECT	TRUNCATE	X
D	SEQ	1	CONFORM	ISOCHORE	FIRST	X
C	SEQ	1	CONFORM	ISOCHORE	FIRST	X
B	SEQ	1	CONFORM	ISOCHORE	FIRST	X
A	SEQ	1	CONTROL	DIRECT	FIRST	X

If the interval-thickness method is used, restore the missing elevations in Figure 12.2 by

elevation(B) = elevation(A) + (A,B)
elevation(C) = elevation(B) + (B,C)
elevation(D) = elevation(C) + (C,D)

These replace Nulls in the appropriate B, C, and D elevation fields.

If the estimated-elevation method is used, the elevation estimates are immediately available, replacing Null values in the field that contains observed picks.

Generate the grids in this sequence by using the estimates of the horizon picks that are combined with the observed elevations. Following the framework table (Table 12.1), grids for the sequence A-D are generated with horizon A as control and horizons B-D conformable to A. These conformables are generated with the isochore method described in chapter 6; horizon B is created by gridding the thickness of the (A,B) interval and adding that isochore grid to the structural grid of A. Next make grid C by calculating the (B,C) thickness at each well, gridding these thicknesses, and adding that grid to grid B. Continue the process for each horizon in the sequence. Because the picks for the eroded horizons have been reconstructed, the resulting grids will show the form before truncation.

If restored data were recorded in the form of thicknesses, some duplication of effort might have resulted, as we first convert these to elevations, and then the isochore method reconverts to thickness for gridding. With forethought, you can reduce time by collecting data by the interval-thickness method, as the interval and observed thicknesses could be combined and thickness gridded directly for building the sequence grids.

If the data were recorded with the thickness-to-reference method—that is, as

(A,B), (A,C), and (A,D)—they can easily be converted to interval-thickness form by subtraction:

$$(B,C) = (A,C) - (A,B)$$
$$(C,D) = (A,D) - (A,C)$$

and then used as above. It is also possible to calculate all conformables from the same control horizon, although this is usually not as desirable. In this case, the thickness-to-reference method could be used directly and interval-thickness values converted.

The next step, as with construction of any stratigraphic framework, is to relate the next-younger horizon or sequence to the grids just constructed. In Figure 12.2, unconformity E is gridded normally and then used to truncate the deeper horizons. Because any of the deeper horizons could be affected, E must be operated against all grids A through D.

The truncation operation can result in either of two types of grids: The "eroded" grid can be set equal in elevation to the truncating horizon (chap. 6), or "eroded" nodes can be set to Null values (chap. 7). Here the eroded part of the surface must be made equal to the truncating horizon, thereby defining a continuous truncated grid. These continuous grids are used for further construction of the framework and to draw cross sections and burial history curves.

Operations are simplified if a shallower sequence laps onto an unconformity. In this case, the lapping grids need only be operated against the unconformity—that is, the grid at the top of the deeper sequence. The baselap operation that creates a continuous grid should also be used here.

The process is complicated slightly for multiple periods of erosion, as described in Table 12.2 and Figure 12.4. Here unconformity horizon D has been subsequently eroded by G. In this case, when D is gridded it is necessary that reconstructed elevations for D be used in the eroded region. Further, reconstructed grid D must be used to truncate horizons A-C. Only at the last step, when the grid of horizon G truncates the older surfaces, does the grid of D attain the present form of horizon D. The process thus follows nature, grids for older surfaces being created before those for younger ones.

Other Considerations

Reconstruction is merely a case of employing tools that have already been developed. These procedures will handle many of the reconstruction problems one is likely to face. There is no single correct way to reconstruct the section, but a variety of possible approaches may be used in conjunction with geologic principles and logic. Now let us consider other aspects of reconstruction.

TABLE 12.2
Framework Table for Figure 12.4

Surface name	Surface type	Sequence number	Sequence control	Initial grid-building method	Primary baselap/ truncate	Special baselap/ truncate
G	UNCF	X	X	DIRECT	TRUNCATE	X
F	SEQ	2	CONFORM	ISOCHORE	BASELAP	X
E	SEQ	2	CONTROL	DIRECT	BASELAP	X
D	UNCF	X	X	DIRECT	TRUNCATE	X
C	SEQ	1	CONFORM	ISOCHORE	FIRST	X
B	SEQ	1	CONFORM	ISOCHORE	FIRST	X
A	SEQ	1	CONTROL	DIRECT	FIRST	X

The previous discussion did not include faulting, a substantial complication, but similar principles apply. First, determine at which stage of the history faulting occurred. For example, if in Figure 12.1 horizons A, B, and C were displaced but horizon D was not, then we conclude that faulting took place after deposition of horizon C but before truncation formed horizon D. The final grids therefore must include noneroded, nonfaulted versions of horizons A-C, plus noneroded, faulted versions of these horizons, as well as their present configuration. If faulting and erosion are interpreted as essentially simultaneous, then a single set of restored grids may be adequate.

Creating the grids is easiest with the restored-surface method of handling faults (cf. chap. 8). In this case, the restored surface that results can be used as the prefaulted configuration. For the fault-block method, we normally begin with a separate set of grids for each fault block. If they had been combined into a single grid, another grid of vertical displacement must then be created. This would probably also be made by use of polygons and would be such that addition of the grid to (or subtraction from) the faulted grid would restore it to a prefaulted condition. This grid could then be applied to other horizons affected by the faults to restore their grids to a nonfaulted configuration, assuming no growth on the faults.

For dipping fault planes where the baselap-truncation method is used, the process becomes much more complicated. Here the geologist is virtually forced to interpret and grid the horizons separately. The faulted horizons are gridded according to normal procedures (other than accounting for postfaulting erosional restoration) by fault block, as are the continuous nonfaulted horizons. However, the nonfaulted data does not exist, so picks must be made for the horizons with fault throw removed—that is, the prefaulted elevations as referenced to the present datum. These picks are then gridded to create the prefaulted horizons.

Regardless of the method used for creating the fault grids, truncation or baselap with the faulted framework must be done. The complete sequence of events then

264

is as follows: (1) Create the nonfaulted grids, restoring eroded thickness as needed; (2) create the faulted grids, again with the eroded thicknesses restored; (3) perform truncation or baselap operations on the grids of Steps 1 and 2 to give eroded and faulted configuration; and (4) continue to the next-younger sequence.

A second consideration involves adequate representation of the thickness grids used in generating conformable surfaces. Low-angle fault planes combined with conformable horizons can lead to poor thickness grids. If the thickness grids (whether restored or of present configuration) do not behave correctly, the procedures described in chapter 9 to force specified interpretations or zero-line contours may be appropriate.

A third consideration concerns compaction of sediments. All the grids that are made with the procedures discussed above are based on present depths and thicknesses. However, the intervals were thicker when deposited, particularly if shale is abundant, and the intervals were compacted to their present thickness as a result of loading over time. If compaction is significant to the problem at hand, "decompacted" grids should be created. There are programs that can calculate compaction (cf. Perrier and Quiblier, 1974; Horowitz, 1976; Bishop, 1979), but these are beyond the scope of this book.

A fourth consideration is wells that are not drilled to all horizons. In this case, the total depth (TD) defines a limit, and all deeper presently observed horizons must have an elevation BELOW the TD (see chap. 5). In this case, use the iterative procedure for gridding when limits are present.

Thickness data involve a similar consideration. In Figure 12.2, the zones are extensively eroded at well 2. If an estimate of (A,B) cannot be made, then the grid at that location might project to small values. However, the gridded thickness must be greater than the observed thickness from unconformity top E to stratigraphic top A. Similarly, the gridded thicknesses (B,C) and (C,D) must be greater than zero. These cases should be treated as limits, as described in chapters 5 and 10.

SHOWING THE RECONSTRUCTED HISTORY

Special Displays

Once the grids have been created, they are used for plotting or making calculations. Displays that might be of interest can range from simple maps of present surfaces, to maps of reconstructed horizons, to cross sections. Concepts discussed in chapter 7 are extended here.

Isochore maps are useful for interpretation (cf. Bishop, 1960; Levorsen, 1967). They can be used to indicate depositional or erosional centers; for instance, Maslyn and Phillips (1984) discuss paleotopographic petroleum traps detected by mapping. Lee (1955) also points out that thickness maps can indicate areas of structural movement. Recall that caution must be used if the thickness interval contains a sequence boundary.

For mapping the present-day interval between two conformable horizons, the procedures discussed in chapter 7 apply here. Isochore mapping of reconstructed intervals is similar. For pre-erosional thickness of a conformable interval, the restored thickness grid used for generating structure would be used. For nonconformable surfaces, subtracting pretruncated or prebaselapped structure grids gives the interval thickness.

Using these methods, it is possible to draw an isochore map corresponding to any geologic time. This requires first selecting those grids that define the appropriate age interval. If there have been multiple periods of erosion, we have a choice of several grids, depending on the stage of reconstruction—that is, before, after, or between specified erosional events. Subtraction of the selected grids and contouring gives the isochore map.

A structure map showing the configuration of a given horizon at some time in the past may be useful (e.g., Appleby et al., 1984). To make this map, first convert the grid of the horizon from its present shape to the configuration it had at the specified time. The conversion is done by using another grid as a reference datum; call this grid B and the grid to be mapped A, as shown in cross section in Figure 12.5A. Restored or pretruncated grids should be used. For the moment, assume that horizon B was horizontal when deposited at the time of interest.

Recalling that the grids are in elevation, the structure in horizon A that existed when horizon B was at the surface is computed as grid $AA = A - B$. Grid $BB = B - B$, consisting of zeros throughout, could also be made. Figure 12.5B shows a cross

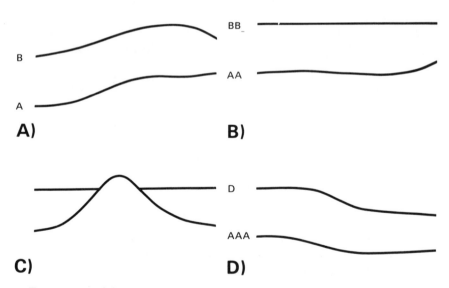

Figure 12.5 *(A)* Cross section showing horizons A and B. *(B)* Section showing horizons adjusted to have configuration at the time B was horizontal. *(C)* Section showing nonconformable adjusted horizons. *(D)* Section showing horizons adjusted for paleobathymetry.

section through grids AA and BB. If B laps onto A and baselap operations are not applied, the restored grid AA might project above BB, indicating an ancient topographic high (Fig. 12.5C).

If reference horizon B was not flat at the time of deposition, but had significant bathymetric expression, then a simple horizontal plane might not be an adequate representation. Assume paleobathymetry is gridded as D. This is commonly a simple surface (e.g., a low-order trend surface), but may be gridded from data or digitized from a hand-drawn map. Another operation forms the paleostructure grid; if D is in terms of elevation, AAA = AA + D. Figure 12.5D shows the surface adjusted by D.

The paleostructure map is obtained by plotting grid AAA or grid AA. If A has no baselapping or truncating relationships, simple contouring will suffice. Paleosubcrop maps, needed if A intersects other surfaces, are made as discussed in chapter 7. However, now the intersecting structural surfaces, not just horizon A, may need adjustment to the reference datum. However, the subcrop lines will not move and may be generated from grids corresponding to the same period of time.

Cross sections corresponding to some ancient time are made similarly to paleostructure maps. All grids representing surfaces older than that of the reference are subtracted from the reference, with paleobathymetry incorporated as needed. Cross sections are plotted normally. Each grid that existed at the reference time is plotted. These grids can be pre- or posterosional, depending on the time of truncation relative to that of the datum.

A series of cross sections, each referenced to a different datum and thus representing a different geologic age, shows structural changes through time and can be valuable in studying geologic development. A series of structure and isochore maps corresponding to several geologic ages also depicts the history. Several authors discuss these concepts (e.g., Wilson and Stearns, 1963; Paynter, 1970; Williams, 1984), but most have used manually drawn maps and sections.

Sometimes a map or section must be referenced to a geologic time that is not represented in the set of horizons and grids. Suppose surface B represents an age of 50 million years (m.y.), and surface A an age of 75 m.y., but we wish to reference a section at 60 m.y. (Fig. 12.6A). If A and B are conformable, linear interpolation on the (A,B) isochore grid, I, might be acceptable. Here a new grid, N, is created by

$$N = B - (60-50)/(75-50) \, I$$
$$= B - 0.4 \, I$$

This grid is then used as a datum to adjust the older grids.

The process is more complicated if A and B are not conformable. For instance, suppose horizon B laps onto A, as in Figure 12.6B. If the depositional rate, r (say, in feet per m.y.), can be estimated for this time interval, generate a grid with every node value equal to $(60 - 50)r = 10r$. Subtracting this grid from the prebaselapped grid B gives the new reference N. Similar considerations apply to truncation if erosional rates can be estimated.

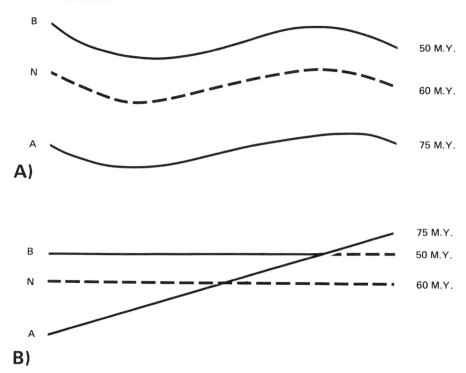

Figure 12.6 Estimation of time horizon where data were not present. *(A)* Linear interpolation by age between conformable horizons. *(B)* Subtracting thickness based on sedimentation rate, *r*, from prebaselapped horizon B.

Burial History Curves

Van Hinte (1978) points out that knowledge of deposition, erosion, and subsidence rates is useful for various aspects of exploration, and he discusses the construction and use of the burial history curve (BHC) in geohistory analysis. The BHC, which graphically shows this history, is a plot of the depth at which a specified point is buried versus geologic time.

The BHC is also useful in tracing thermal and pressure histories (Guidish et al., 1985). These depth-time relations and the implied overburden pressures allow prediction of reservoir porosity, compaction, and temperature (e.g., Rittenhouse, 1971a, 1971b; Stephenson, 1977; Houseknecht, 1984; Schmoker, 1984). Temperature-depth gradients and/or heat flow can be used to calculate temperature history (e.g., Cercone, 1983; Hitchon, 1984). Time-temperature can be related to chemical reaction rates, allowing study of diagenesis (Siever, 1983) or the prediction of hydrocarbon maturation or quality (e.g., Waples, 1980; Cohen, 1981; Middleton and Falvey, 1983; Hitchon, 1984; Tissot, 1984; Furlong and Edman, 1984). Such

TABLE 12.3
Geologic History of the Area Shown in Figure 12.1

Horizon	Time (m.y. before present)	Event
A	50	Deposited
B	40	Deposition conformable to A
C	30	Deposition conformable to B
C′	20	Deposition conformable to C; not seen in area due to subsequent erosion
D	15	Uplift, folding, and erosion to form unconformity
E	10	Deposition lapping onto D
F	5	Deposition conformable to E
F′	3	Deposition conformable to F; not seen in area due to subsequent erosion
G	0	Uplift and erosion to form unconformity; present surface

calculations allow the geologist to relate porosity and hydrocarbon quality to migration and trapping.

A burial history curve is constructed for a specified point on a given horizon. Consider the example cross section of Figure 12.1. The history and ages of the horizons are summarized in Table 12.3. Figure 12.7 shows the BHC that corresponds to the oldest horizon (A) at the location of the well. We therefore study the history of that piece of rock where the borehole intersects horizon A.

In Figure 12.7, point A represents deposition of the specified rock at the surface (depth = 0) 50 m.y. ago. Deposition continued until horizon B was created 40 m.y. ago. At this time horizon B was at the surface, and the rock was buried at the depth equaling thickness (A,B), indicated by point B on the plot. Continued deposition through C leads to C′, representing the maximum depth at which the rock has been buried so far.

Horizon C′, having been eroded, does not presently exist anywhere in this area, so it would be necessary to reconstruct it everywhere. For normal analysis as discussed above, it is not necessary to interpret an elevation for C′. However, knowledge of C′ is needed here to show the maximum extent of burial.

Uplift and erosion occur after C′ is deposited, and some of the overburden is removed. Point D represents the depth after unconformity D was created by erosion, and the upward trend of the curve from C′ to D reflects this reduction in depth. Further deposition from 15 to 3 m.y. ago buries the rock deeper to F′, another presently nonexistent horizon. Finally, erosion forms the present topographic surface, G, and decreases the depth of burial to its current amount.

If BHCs are required at a number of locations, as is commonly the case,

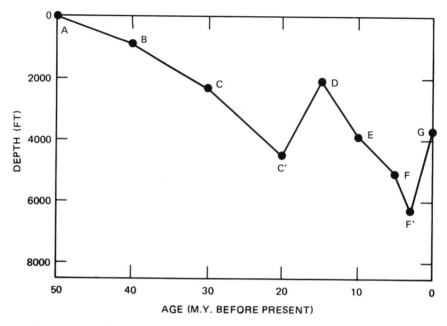

Figure 12.7 Burial history curve corresponding to history in Table 12.3 and Figure 12.1. This curve plots the burial history for the deepest horizon, A, at the location of the well.

calculations can become tedious. However, a set of grids that depict geologic history can be used. A grid representing the depth of burial of horizon A can be created for each grid corresponding to a younger horizon. Creation of these grids follows a process similar to that used for generating a cross section referenced to a given horizon, though with signs reversed. For instance, the depth corresponding to deposition of horizon B is given by subtraction: B − A. Values can be interpolated from these depth grids at each of the locations of interest. These interpolated depths, plus the corresponding ages, are used to plot the BHCs for each location.

An alternative but equivalent procedure is to interpolate a value at every location of interest from each of the structural grids in the restored framework. Depending on the stage of the geologic history, these structure grids may be pre- or post-truncated, -baselapped, or -restored. Subtraction of the interpolated elevation of horizon A from each of the other values gives the series of burial depths.

Another aspect must be taken into account when constructing BHCs. To be complete, bathymetry should be included. Depth of burial refers to thickness of rock overburden — that is, it is calculated from the grass roots or sea bottom of the current age. However, showing the effects of changing bathymetry on the plot can also be valuable (cf. van Hinte, 1978). In any event, because the BHC must extend to the present, it is necessary that a grid of the present topographic or bathymetric surface be used.

270

To be strictly correct, BHCs should include the effects of sediment compaction. If this is important for the area being studied, grids should be generated that show thickness before, during, and after compaction. Perrier and Quiblier (1974), Horowitz (1976), and Bishop (1979) give methods for estimating compaction effects. Brunet (1984) discusses development and application of subsidence curves.

Suppose we wish to construct a second BHC at the location of the well in Figure 12.1, but now for horizon B. This curve will start at the time of B and will parallel the curve for horizon A. The two curves will be separated vertically by (A,B) throughout. It is also possible to combine curves. If we model hydrocarbon maturation in a source rock at horizon A, the history will follow the curve in Figure 12.7. Vertical migration of the hydrocarbon to reservoir rock at horizon B causes the BHC to jump to the B-curve.

Afterword

This book has stressed the importance of placing geological interpretation into grids used to create contour maps. This is done through the use of three important factors: geological reasoning, a problem-solving approach or philosophy to mapping, and a set of procedures or tools. The first two of these factors are used in concert to determine the significance of geological properties and to relate those properties to mapping. The set of procedures—that is, the bag of tools and tricks—is used to create the maps.

A good grasp of these three factors, plus access to a modern mapping system, allows consistent, realistic maps to be generated. In addition, there are many applications beyond the discussion in this book. The methods and examples presented have only scratched the surface. Although the basic approach to problem solving is widely useful, every special geologic situation requires different uses of the tools. In addition, some problems require special geologic assumptions or reasoning.

The following example shows how tools can be modified to solve mapping problems. The thickness of a coal bed may be mapped routinely. However, the bed may contain partings (shale breaks) that reduce the volume of coal. These partings may be erratic in lateral extent and difficult or impossible to correlate. If we merely cumulate the total shale thickness at each drillhole and grid, the thickness of shale will often be too great between holes. Subtraction of shale thickness from the coal isochore could thus give an erroneous net coal isochore.

If correlation of the shale breaks is impossible, perhaps we might relate them arbitrarily from hole to hole. For instance, a simple method would assume that the top shale break in a hole correlates with the top break in every other hole, the second break in each hole is correlated, and so on; a similar procedure could start at the base of the coal and work upward.

The following steps could be used to generate a grid of thickness of a given shale (say, the first parting).

1. Create an isochore grid of the "correlated" shale break; where shale does not exist in a drillhole, use a Null value in the data (i.e., allow thickness to project through that location).
2. Using shale-thickness data, create a new data field; set values where shale exists to 2 and values where shale is absent to −1.
3. Create an "indicator" grid from the (−1, 2) data.
4. Modify the grid nodes in this indicator grid as follows: Set all nodes greater than 1 to 1, and all negative nodes to zero.
5. Multiply this modified grid by the shale isochore grid; the resulting isochore grid will retain shale where present, remove projected thickness where absent, and show shale pinch-outs between drillholes.

The total shale-thickness grid is obtained by adding individual thickness grids.

Although the tools and approaches described in this book can be used for most projects, modification is often necessary to fit the specific situation. The correlation of first shale breaks in order to build a grid of parting thickness is an approach that, although not based on geology, was necessary to calculate coal volumes. The indicator grid is similar to that used for defining zero contours. Because drilling density in mining projects is often high, the indicator technique usually works well. The (−1, 2) modification was needed to produce consistent values of zero and 1 in the modified grid (Step 4); otherwise, grid values could have fluctuated, sometimes above 1 and below zero. This example shows that pragmatic approaches are sometimes necessary, with methods perhaps not strictly geology-based. Willingness to experiment and blend various techniques is important and useful in computer mapping.

This book has concentrated on mapping geologic surfaces. However, many other types of data involve rock attributes. Aside from a brief discussion as part of volumetrics calculations, time and space considerations prevent us from considering such variables, although their applications are numerous enough to warrant a second book.

Rocks have many mappable attributes. Maps can be made of such sedimentary properties as porosity, permeability, grain size, sand-shale ratios, and facies. Petrology and geochemistry may require mapping oxide abundances, ternary variables, ratios, or mixtures of populations. Paleontologic maps may describe fossil assemblages. Even orientation data (e.g., cross-bed directions, poles to mineral grains, or dip of bedding) are mapped.

Porosity is a commonly mapped attribute in the petroleum industry. Average porosity over an interval can be mapped directly; however, reservoir engineering calculations require porosity thickness. A porosity-thickness grid could be made in two ways: calculating grids of thickness and average porosity separately and multiplying them, or calculating porosity thickness at each well and gridding the values. The same answer will be obtained at the data points, but the maps will vary between points.

Similar considerations apply to mapping ratios; typical examples include sand-shale ratios, percent sand in an interval, ternary variables, and other facies properties. In addition, all of these examples must deal with problems of sparse data, positioning zero-lines for each variable, and restriction to maximum or minimum allowable limits.

Other considerations involve mapping variables with specific statistical properties (e.g., lognormal distribution) or variables that are a mixture of several populations. For instance, a geochemical survey may show background with a few anomalies. If the anomalies must be carefully defined, it may be necessary to separate them from background, map them individually, and recombine them with the background later.

These examples indicate just a few of the varied types of problems associated with mapping rock-attribute data. However, such variables can also be mapped through use of the procedures in this book. Some tools are broadly applicable, some can be used with minor modification, and others need to be developed for special cases. The general approach to problem solving is useful, as we find that virtually every geologic situation is slightly different.

We hope that readers are inspired to use the tools presented in this book and to develop mapping methods that go beyond what we present. Further, it is likely that experienced mappers have developed tools that are not discussed here. It is our hope that this book spurs the computer-mapping community to share procedures and tools through journal publications and presentations at conventions.

275

APPENDIX A

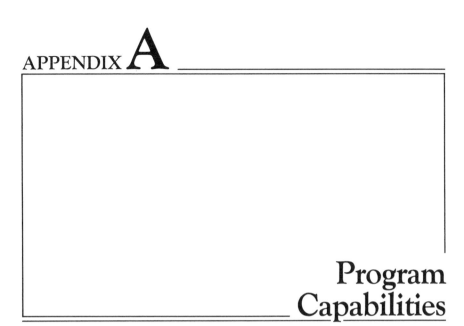

Program
Capabilities

In order to follow the procedures outlined in this book, the mapping program must have certain capabilities. This appendix summarizes abilities that are necessary for most projects; describes some useful, optional capabilities; and provides a useful list of requirements for constructing or purchasing a program. Walters (1969) and Waters (1981) also describe general characteristics of contouring programs.

DATA ENTRY

The program must be able to put data into the computer and then organize it for use by the program options. Descriptions of five important data-entry capabilities follow. Certain abilities in addition to simple data entry are also required, either in the mapping program or in some external program. These are discussed under data calculation.

1. The program should be able to read, use, and pass to other options several variables or data fields. It should be possible to input one or more Z-fields in addition to the X and Y coordinates. (The Z-variables are the values observed at the (X,Y) location.) It should be possible to specify formats describing the location and form of the data on the input records. In writing a program, the number of available fields should be generous; even small projects can require many fields or variables, and commonly many new fields are created from the input fields (e.g., calculation of thickness from structural elevations). An upper limit of 100 fields is not excessive.

277

2. Most Z-fields contain numeric data, such as elevation, thickness, mineral content, and so on. However, some fields will need to handle alphabetic data (for example, well or sample identification) for posting on base maps or to contain coded information. It is important that the user can specify which fields are alphabetic and that the fields be read as such.

3. The program should be able to combine data files for corresponding samples, or to concatentate files—that is, to add a file to the end of another.

4. The program should have the capability to detect a blank field (one in which no value is given) and automatically insert a Null value (see below). Many FORTRAN programs convert a blank field to a zero value if read numerically; the program should be able (optionally) to check for and correct this.

5. The program should be able to specify minimum or maximum limits on coordinates of the input data. Points outside the specified area would not be included in further processing. Similarly, polygons might be used to specify areas to contain (or exclude) data.

NULL VALUES

Many datasets contain missing values because of, for example, measurement problems, physical or time constraints, or inability to observe a horizon (e.g., rock unit missing or well not drilled deeply enough). A gap in a dataset causes difficulties, so the computer must be instructed that a given entry is missing. This is done through use of a Null value, a number that realistically cannot occur (e.g., -10^{30}) but that holds a position in the data array.

Whenever the program detects a Null value in a set of data, the computer must be instructed to treat it specially and not use it for creating grids. The Null value should also have special handling when values between various fields in a set of data are calculated. For example, when subtracting two structural elevations to obtain thickness, if one of the elevations is a Null value, the thickness is undefined and should be set to the Null value.

Null values can also occur in grids (see below). Gridding methods may calculate a value at every intersection, so the grid would have no missing values. However, restricting the gridded portion (e.g., by distance to data or by polygons) is common, and Nulls are used at the blanked intersections. For example, we might convert the intersection values to Nulls in a region that is not to be contoured; the program would leave Null regions blank. Similarly, when one grid is operated against another, Nulls define areas for no calculation or other special treatment.

DATA CALCULATIONS AND MODIFICATION

It is often necessary in a project to modify the input data or to perform calculations between the fields. For instance, consider a dataset containing elevation fields for two horizons. Subtraction of one elevation from the other (for each well or data point) provides a new field of thickness. This procedure would require that the user can specify (1) the operation (subtraction), (2) identification (numbers or code names) of the fields to be operated on, and (3) where the result is to be placed (one of the original fields or create a new field). If one or both values are Nulls, then the result should be Null.

Many operations in this data-calculation step may be valuable, and even obscure operations might have some use. The available calculations should be both arithmetic and logical. Arithmetic operations include adding, subtracting, multiplying, or dividing the values in the two fields. Similarly, such operations as adding or multiplying the values in a field by a constant can be useful. Other arithmetic operations include square roots, taking logarithms, raising values to exponents, and trigonometric functions.

Logical operations involve taking certain actions, depending on values in the two fields. For example, we might wish to retain the greater or the smaller of the values in two fields at a data location, or we may instruct the program to delete an entire data record if a value in a given field meets some test criterion.

Some operations work with Null values. For instance, we may wish to change a value to Null, thereby effectively eliminating it, if the value in a given field is greater than some specified constant cutoff; this can be useful for creating a subset of the original data if a field includes identifiers or codes. We may also wish to replace Null values by constants.

An important calculation, missing from some widely used programs, is necessary at the end of Data-merge procedures (chaps. 5 and 6). Suppose we have a field, A, that contains both observed values and Null values, and a second field, B, that contains calculated values from Data-merge. The calculation must then replace Null values in field A with the corresponding values from field B, but all real values in A must be left unchanged.

This calculation for Data-merge points out the need to perform calculations or modifications at any stage of processing. Most such calculations are done at data entry, but the ability should exist to do them later in the job stream.

BASE MAP

When the data are first analyzed, a base map is valuable for determining well locations and spacing, finding trends and relationships between variables or horizons, and so on. While it is not directly involved with contouring, we recommend that a good base-mapping capability be included in the program.

A base-mapping option should be able to post data from several fields at each data location, ideally with control on positioning around the point. The map's X-Y limits, scale, and titles should be under user control, and it should be possible to overplot the base map on contour maps, and vice versa. Other capabilities might use polygons to control posting, or plot text, lines, or polygons on the map. A valuable option for petroleum applications is the ability to post different symbols on the map, corresponding to a code in one of the data fields.

GRIDS OR GRID EQUIVALENTS

For contour mapping, surface manipulation, volumetric calculations, and other tasks, a regularly distributed set of points is easier for the program to work with than a random set. A rectangular array of values is used by most large programs, although some use a triangular mesh. In either case, values from the original, irregularly spaced data are assigned to each array point, so the array contains a complete set of information about the surface. Grids are discussed in detail in chapter 4.

The grid used by most programs is a rectangular array of numbers—that is, a series of rows and columns. The spacing between rows in the array is generally constant, as is the spacing between columns, although the row and column spacings may differ. The rows and columns meet at intersections, or nodal points. A framework is thus defined over the area being mapped.

Suppose we are mapping the structural elevation of some surface. If we knew the elevation at each of the grid intersections, we would have detailed knowledge of the form of the surface and would not need to use the well data. Gridding algorithms interpolate the scattered well information and assign a value to each grid intersection. Grid creation thus consists of two steps: (1) definition of the framework and (2) assignment of values to the nodes.

The grid that results is used for contouring, cross sections, operations with other grids, integration, and so on. Most projects require more than a single grid, so the user must be able to define a consistent set of X-Y limits and grid interval—that is, the program's algorithm should not set these without control. A single set of limits and intervals should be used throughout the project. It is also necessary that the grids be identified with numbers, code names, or locations in a file for later access.

The grid need not be rectangular or square; a rectangular X-Y grid limit can be defined, but the inner mesh could be made up of triangles, usually with a data point at each vertex. If a deep horizon has fewer data points than a shallow one, it may be necessary that points corresponding to the shallow points be added to the deeper data. If so, the triangle mesh would be defined with all these points, plus interpolated values for the added points from the surrounding data. Triangular grids have advantages, notably that their maps honor the data. However, they can require special processing for baselap, truncation, or defining zero-line locations.

It is not strictly necessary that grids be used. It is possible to work with a grid-equivalent database. This might take the form of a set of data, rather than the specific format of a grid. The data locations commonly would be in the form of an array. Values must be assigned from the wells to the database so defined, as with a grid. However, the database allows a number of Z-values to be assigned to each X-Y location. These values can then be calculated or operated one against the other, as with fields in a dataset. The program would need the ability to contour, integrate, and so on from this grid equivalent.

GRID CREATION

The program must be able to use irregularly spaced, linear, or digitized data to create a grid. Virtually all programs used for mapping will create grids, whether square, rectangular, or triangular, or some other equivalent. General methods for building grids are discussed in chapter 4.

Important user controls include the grid interval, restrictions on points selected, and calculation methods. Points to be used at a node can be selected on the basis of (1) distance and direction from the node, with points beyond some cutoff distance not used; (2) search sectors, ensuring that points are evenly distributed around the node rather than being clustered; and (3) maximum or minimum number of points to be used at the node. Along with a choice of calculation methods, weights used to discount distant points must be selected.

Two other capabilities are less important. The first restricts gridding calculations to a neighborhood of the data points (see chap. 4), preventing extrapolations. The ability to calculate only intersections within a convex hull surrounding the data points, or within an even more restricted region, is useful; the outer intersections would be assigned Null values. A similar restriction assigns a Null to a given grid intersection if it exceeds a limiting distance to the nearest data point. Polygons can also be used to restrict gridding. Some programs allow a second grid to define which nodes are assigned values, depending on the values of its corresponding nodes.

The second option is the ability to specify that the value of a data point falling directly on a grid intersection is to be assigned directly to the intersection. Two advantages result if a dataset that came from a grid (e.g., grid-to-data) has such values assigned directly: reduced computer time in gridding, and assurance that the grid is replicated exactly in the region of the grid-data.

EXECUTIVE SYSTEM AND GRID FILE

It is possible to carry out the methods described in this book with a series of separate programs: one program for data calculation, another to create a grid, a third for contouring, and so on. However, the task of processing a complicated framework would be tedious and time consuming. Similarly, separate files that hold only single grids would be difficult to use. We accordingly recommend using a program that contains an executive system and integrated grid file.

An executive system consists of a main routine that can receive such general task commands as **CREATE GRID, PLOT CONTOURS,** or **INTEGRATE.** The program reads the command (whether in words, an abbreviated code, or

281

interactive screen input), interprets the meaning, and directs control to the appropriate part of the program. This program task can read and execute the specific detailed controls. After this task is completed, the program returns to the main executive.

Executive-type systems can be designed in a number of different ways, and it is difficult to judge one better than another. They may be highly formalized, as is the statistical analysis system described by Jones et al. (1976), or simple in form and concept. The important factor is that the user can easily control program flow from one option to the next. Most major mapping programs, whether organized for interactive or batch processing, have this ability.

Along with an executive system, an integrated grid file is important. Each option or program task should be able to obtain the needed grid from a standard file and add the resulting grids to the file. The grids should be identifiable for later use with some code number or name. Jones et al. (1976) describe a program that handles statistical datasets on a file similarly. Again, their system is more complex than strictly required. In addition, the program should use internal data, polygons, or other files that can be passed from option to option.

PLOTTER CONTOURING

Once grids are created, they are normally used to plot contour maps. Standard programs should include the ability to set the scale, X-Y limits, contour interval, title, and contour-line widths and annotations. Five other abilities are needed to provide flexibility in mapping, of which the first two are very important.

1. The program should be able to overplot contours onto a map previously drawn from another grid. For example, the construction of subcrop maps (chap. 7) requires that structure contours from one grid be drawn and the subcrop line (from a thickness grid) be overplotted on the first map. In addition, it is useful to overplot a contour map onto a base map.

2. It is often necessary to restrict the values of the contours that are to be plotted. For example, thickness grids should include negative projections. We do not wish to see the negative contour lines when thickness grids are plotted, so the program should be instructed to draw only contours corresponding to values of zero and greater.

3. Areal restrictions on contours can be valuable. Areas of a grid that contain Null values are normally left blank. However, areas could be defined with polygons to specify regions that are to be masked.

4. The program should be able to plot special contours. For example, the outline of a horizontal gas-oil contact may be needed on a structure map. The elevation of this contour may differ from any of the usual contour values, so a special contour having the appropriate value would be plotted.

5. Chapter 9 shows that it can be useful for a contouring program to digitize contours automatically and write them onto a file. This information consists of the

282

X-Y coordinates of points on the contour, commonly where the contour crosses grid rows or columns, and should be in a form (polygons, line segments, or special format) that other program options can use.

CALCULATIONS ON OR BETWEEN GRIDS

It is rare that a final grid, ready for contouring, comes directly from a grid-creation step. Much more commonly we must manipulate grids one against the other and use the result of several operations (cf. chap. 6). This concept is similar to operations between fields in a dataset.

Grid operations can use one or two grids. Single-grid operations involve a simple calculation on each intersection value. For instance, a thickness grid is converted from feet to meters by dividing every grid-node value by 3.28. Other calculations include arithmetic operations, applying trigonometric functions, taking square roots, and so on at each intersection. Another useful ability creates a grid with every node assigned the same given value.

Operations can also involve two grids. Recall that each intersection in a grid corresponds to an X-Y location. Hence, if several grids cover the same area, corresponding intersections (e.g., row 1, column 1) represent the same X-Y location, and a calculation can be made using the values from the matching intersections. For instance, suppose we have two grids of structural elevation. To create a grid of thickness, we perform the following steps:

1. Relate a given intersection in one grid to the corresponding intersection in the second;
2. Determine the Z-value associated with each intersection;
3. Subtract one Z-value from the other to give thickness at the corresponding intersection in the new grid;
4. Repeat this process for every intersection to create the grid.

The resulting grid should then be available for contouring, further operations, and so on.

We must also be able to add (add values of corresponding intersections), multiply, and divide grids. Other similar operations might consist of averaging the values of two grids, calculating directions from sine and cosine grids, and the like.

For some uses, it is also necessary that minimum or maximum limits be specified on the input grid values. A good example is the addition of a thickness grid to a structure grid to obtain another structure. If the thickness grid projects to negative values, the grid intersections must be changed to zeros before addition. The ability to specify a minimum of zero allows this to be accomplished. A similar capability on output grids is also useful.

Logical operations, particularly two stratigraphic operations, are necessary (chap. 6). Truncation is an operation in which the intersection values in two grids are compared. Let B represent a grid that is to be truncated by grid A, and assume the Z-values are in terms of elevation. To create the new truncated grid BB, copy unchanged the B values deeper than A, but replace the values of B that are greater than values from the A-grid with values from A. The truncation operation thus outputs the lesser (deeper) of the elevations:

$$BB = minimum(A, B).$$

The second important stratigraphic operation laps a grid onto an older surface. If grid A is to be lapped onto B, create a new grid AA that contains the greater (shallower) of the values from the A and B intersections, as

$$AA = maximum(A, B).$$

In this case the AA grid is identical to A where it is above B and coincident to B where A has projected through the older B surface.

These two stratigraphic operations create continuous grids, which are necessary for generating a framework. However, subcrop maps require different grids from those resulting from the truncation and baselap operations. For truncation, retain the untruncated portion unchanged (as with the continuous grid just discussed), but blank out the truncated intersections (i.e., where values of B are greater than A) by converting them to Null values. If this grid is contoured, it will contain both contours and blank areas. A similar blanking operation should be available for the case of lapping a surface onto an older surface.

Other useful operations include restricting the input grid values with minimum or maximum limits and then testing versus real or Null values. As with calculations on data, a choice of many operations will aid in doing large projects.

INVERSE INTERPOLATION

Interpolation of a value from a grid is an extremely valuable option. Suppose we have a structure grid and wish to know the elevation represented at some location (e.g., a proposed well) not necessarily on a grid intersection. The program must find those intersections near the proposed location and then use the associated nodal values to interpolate a value for the location. Normally 4, 12, or 16 grid intersections are used.

This procedure is meant to serve as an inverse of the gridding process—that is, it goes from a grid to a set of data values. However, in most cases the process is not a true inverse, as the set of interpolated values would not normally reproduce the grid if used for recalculation. Common methods use simple surfaces (polynomials or splines) or interpolations to calculate the value, whereas most gridding algorithms are more complex.

284

Inverse interpolation is used in many of the procedures discussed in this book, notably in chapter 5. We usually must make the calculation at many locations (for instance, at all the original well locations), rather than only at a single data point. These locations are conveniently introduced to the option by reading the file created by the data-entry step. While perhaps not strictly necessary, processing will be simplified if the program can interpolate from several grids at once for each of the X-Y locations.

Output from this option is a new set of data, either in the form of the file generated by the data entry step or as a file of card images. Each data point must include the X-Y coordinate and the interpolated values at that location. In addition, including original data fields from the input file can be useful, as the observed and gridded values can be combined and the specially modified data regridded.

GRID-TO-DATA

Grid-to-data is an option that allows grid information to be converted to data. A single grid containing many Z-values is transformed to many records, each corresponding to an intersection and containing the (X,Y) coordinate and Z-value. All information in the grid is thus converted to a data file. Grid-to-data also should be able to handle several grids at once; then each record of the output file would contain X, Y, and several Z-values.

Consider the following example of how grid-to-data might be used. Assume that a thickness grid has positive values but contains zero or Null values where thickness is absent. For many purposes (see chap. 9), the grid should project to negative values from the gradient of positive values. A new grid can be made with three steps: (1) Convert every grid intersection to a data record containing X, Y, and Z values (grid-to-data); (2) if the grid contains zero values, change them to Nulls so the gridding routine will ignore them; and (3) grid the remaining data. The grid that results should be the same as the original where the grid existed, but will be newly projected in previous Null regions. Of course, the grid-creation routine must be able to project trends or slopes.

CROSS SECTIONS AND PROFILES

The capability to plot cross sections or profiles through grids is an important part of a major mapping program. These are valuable displays for final presentation, but are also useful for checking relationships between grids and subtle forms in individual surfaces. Whenever a complicated series of grid operations is performed, it is wise to check the results at all or some of the intermediate steps. It is difficult to do this with contour maps; structural and stratigraphic relationships normally show up well on cross sections. We recommend that liberal use of cross sections be made to validate individual grids or groups of grids.

The option should have the ability to draw the section between any two locations within the grid limits; ideally it could plot a series of connected segments. In addition, several grids should be drawn on the same plot. User controls include vertical and horizontal scales, titles, and annotations. The ability to project wells onto the section and plot horizon picks is important.

DISPLAYS ON THE PRINTER

The line printer is a valuable tool for obtaining displays of data or grids. While generally not drawn to exact scales or suitable for formal presentations, these displays speed a project because they allow quick evaluations. A simple base map (perhaps with only one or two fields posted and using an approximate scale) allows data distribution to be evaluated with minimum time or cost.

A simple printer map can indicate the form of a surface or grid. Although a program can be designed to draw contours on the printer, a continuous-symbol map is generally more readable, although it is not to scale. In this map, each grid intersection is made to correspond to a printer position on the page. A set of intervals on the Z-variable is defined, with a distinct printer symbol assigned to each, quickly providing a map (see Fig. 4.7). We recommend this procedure for scanning intermediate grids to be sure no errors have crept into processing.

POLYGONS

As discussed in chapters 4 and 8, polygons delimit an area by enclosing it within a series of line segments that connect a set of vertices. The polygon vertices, defined by their X-Y locations, are ordered, so that the first line segment extends from vertex 1 to vertex 2, the second segment extends from vertex 2 to vertex 3, and so on. The last segment extends from the last vertex back to the first, thereby enclosing the area.

Polygons are used in the fault-block method of mapping in the presence of faults. However, they also define or restrict calculation regions, as with the following examples: (1) When creating a grid, calculate and assign intersections only within the polygon, and assign Null values to all intersections outside; (2) when plotting a contour map, plot only those contours that fall within the polygon, and leave the outside blank; (3) when operating on grids, perform the specified calculations only within the polygon, and output Null values or perform some other calculation outside the polygon; and (4) integrate only within the polygon. An added feature is to reverse this procedure, with calculations performed only outside of the polygon.

286

INTEGRATION

Numerical integration of a grid, described in chapter 10, is necessary in order to do volumetrics calculations. The integration option should be able to calculate volumes between a reference datum (normally zero) and the grid surface. It should also be able to calculate projected and surface area. Useful options include: (1) minimum and maximum limiting values specified on the grid, (2) different reference datums, (3) integration between two grids, and (4) integration between the surface and several reference datums (e.g., bench elevations in an open-pit mine).

The ability to read polygons from a file and then integrate up to the boundary they define is very important. These polygons might represent lease boundaries. In addition, a second set of polygons might represent fault blocks. We would need to integrate over the intersection of the two sets of polygons.

Integration near zero-lines can be improved if the program has the ability to read and use special information in the form of a digitized zero-line. This line could come from hand or machine digitization or from the plotter-contouring option.

TREND SURFACES AND FILTERS

Trend analysis options (chap. 11) should allow specifying the required method (polynomial or Fourier) and degree or order. They should use information from the standard data file or from a grid. The function that is fit should be used to generate a grid of the trend through calculation of the function at every node. Ability to define filters and apply them to grids also needs to be included. In addition, filters are used to introduce special smoothing (cf. Briggs, 1974), gradient detection, and so on.

UTILITIES

Utility options are necessary to do a project effectively and easily. These utilities should have the capability to (1) copy or transfer data files from one computer unit (tape or disc) to another, (2) copy or transfer grids from one file to another, (3) list the identifiers and characteristics of grids on the file, (4) print the grid intersection values, and (5) copy or transfer polygons from one file to another. The copy or transfer operations should allow specified portions of the information to be moved, with the rest being deleted.

Other sorts of utilities include the ability to calculate statistical summaries of the information in data files or grids. In addition to analysis of Z-values, data distribution or density is important.

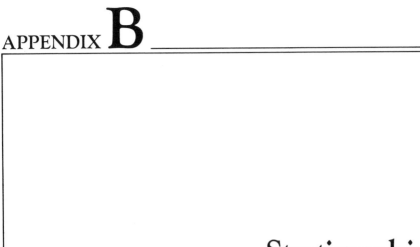

Stratigraphic
Example

This appendix presents a set of data that can be used for creating grids of a stratigraphic framework. The surfaces and their relationships are individually fairly simple, but as a group they include complex features. Our goal is for the reader to use the tools discussed in this book on this data and then compare his maps with ours. General agreement should result, although different gridding algorithms and control parameters can lead to slightly different maps.

Figure B.1 shows a diagrammatic cross section through the area. The Lamont sand is the oldest unit for which data are recorded. Local shale-filled channels occur within the Lamont sand, possibly extending through its base. The top of channeling is recognized as a thin marker on gamma-ray logs. This marker is conformable to the top of the Lamont and divides the Lamont into upper and lower members. The top of the upper Lamont is labeled TULM in Figure B.1. The top of the Upper Lamont is actually an unconformity, but is concordant and lies within the Lamont; it is therefore treated as a conformable surface within the sequence below it. The top and base of the lower Lamont are labeled TLLM and BLLM, respectively. The base of the channel is labeled CHNL.

An unnamed siltstone is found locally above the Lamont sand, and the Aurora Creek sand lies above this siltstone. The top of the siltstone (TSLT) and top of the Aurora Creek (TACK) are conformable to each other, but lap onto the Lamont sand. Shale lies above the Aurora Creek sand. The section is truncated by an unconformity (UNCF) that was developed by erosion of a mature paleodrainage system.

Table B.1 shows the framework table (see chap. 6) corresponding to this area

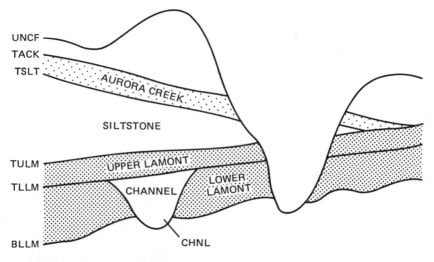

Figure B.1 Diagrammatic cross section showing horizons to be mapped. The two members of the Lamont sand are defined by the top of the upper Lamont (TULM), and the top and base of the lower unit (TLLM and BLLM). The base of the buried channel is CHNL. The top and base of the Aurora Creek sand are defined by TACK and TSLT, respectively. An extensive unconformity (UNCF) truncates the section.

TABLE B.1
Framework Table for Figure B.1

Surface name	Surface type	Sequence number	Sequence control	Initial grid-building method	Primary baselap/ truncate	Special baselap/ truncate
UNCF	UNCF	X	X	DIRECT	TRUNCATE	X
TACK	SEQ	2	CONFORM	ISOCHORE	BASELAP	X
TSLT	SEQ	2	CONTROL	DIRECT	BASELAP	X
TULM	SEQ	1	CONFORM	ISOCHORE	BASELAP	X
TLLM	SEQ	1	CONTROL	DIRECT	BASELAP	TRUNCATE
CHNL	SPEC	1	X	ISOCHORE	X	TRUNCATE
BLLM	UNCF	X	X	DIRECT	FIRST	X

and geologic history. Note that the Lamont and Aurora Creek sandstones each define baselapping sequences. The unconformity and channel truncate sequences below them. The channel is handled as a special surface.

Table B.2 lists elevations recorded in 17 wells for the seven horizons. The fields are labeled according to the codes on Figure B.1. Three additional fields include X

290

TABLE B.2
Well Data Recorded for the Example Study

X	Y	ID		UNCF	TACK	TSLT	TULM	TLLM	CHNL	BLLM
32000	87015	W	1	−7888	−8138	−8139	−8153	−8158		−8210
38600	87030	W	2	−8020	−8058	−8073	−8078	−8086		−8137
44000	86400	W	3	−7943		−7957	−7958	−7970		−8025
48200	87060	W	4	−8003						
30650	82200	W	5	−7918	−8003	−8013	−8038	−8048	−8103	
35030	80901	W	6	−7905		−7943	−7964	−7976		−8027
39476	80970	W	7	−7900		−7928	−7935	−7949		−8004
44198	80910	W	8	−7998						
33530	76230	W	9	−7879		−7901	−7929	−7944	−8004	
38024	76170	W	10	−7817	−7878	−7933	−7959	−7976		−8032
42440	76650	W	11	−8018						
46955	76620	W	12	−7890						
30350	71700	W	13	−7903	−7907	−7961	−7963	−7980		−8033
35195	71745	W	14	−7815	−7946	−7998	−8025	−8045	−8120	
39650	71760	W	15	−7966	−7996	−8043	−8070	−8092		−8153
44180	71766	W	16	−7913		−7925	−7928	−7951		−8028
48800	70215	W	17	−7818	−7920	−7972	−7973	−7999		−8050

and Y coordinates and well identification (ID). Seismic surveys have been conducted in this area, but only the unconformity (UNCF) can be detected. Table B.3 lists elevations of UNCF for 67 shotpoints. The entries in the ID field combine line (letter code) and shotpoint (number).

Because of problems with velocity surveys and line-ties, the seismic and well data are not consistent in their measurements of the elevation of UNCF. Datamerge (see chap. 5) was used to merge the datasets, resulting in the combined data of Table B.4. Note that several of the merged seismic values were eliminated because they were too near wells. This dataset was used to create the map of UNCF in Figure 4.3 and the trend-surface maps in Figure 11.4.

The channel is limited in areal extent, and it is difficult to create a reasonable map of the base of the channel (CHNL) or its thickness. Accordingly, methods described in chapter 9 were used to create nine extra control points of channel thickness (THK). These, shown in Table B.5, were used to aid construction of channel-related grids.

The grids in this stratigraphic framework were created using data in Table B.4 and the procedures described in chapter 6. Figure B.2 shows a cross section through the completed framework, extending from (29000, 75000) to (50000, 75000). Another cross section, extending from well W5 to W10 to W17, is shown in Figure 7.1.

Figure B.3 shows a structure map of the top of the siltstone (TSLT). The blank

291

TABLE B.3
Seismic Data Recorded for the Example Study

X	Y	ID	UNCF	X	Y	ID	UNCF
29240	81960	A610	−7958	37280	71070	D556	−7843
30800	82800	A612	−7967	38660	70020	D558	−7895
32270	83760	A614	−7978	40040	69000	D560	−7927
33770	84720	A616	−7988	41450	67890	D562	−7929
35240	85590	A618	−8011	42800	66900	D564	−7905
36740	86520	A620	−8068	44240	65730	D566	−7868
38210	87480	A622	−8135	45500	64710	D568	−7814
39680	88350	A624	−8165	46940	63600	D570	−7760
41240	89250	A626	−8191	48200	62520	D572	−7720
29840	67560	B068	−7785	49610	61410	D574	−7674
31280	66600	B070	−7757	33800	89220	E003	−8018
32720	65610	B072	−7763	34880	87720	E005	−8023
34190	64650	B074	−7811	35840	86280	E007	−8037
35720	63720	B076	−7889	37040	84720	E010	−8026
37100	62700	B078	−7928	37940	83160	E012	−7989
38600	61800	B080	−7947	38840	81900	E014	−7957
40100	60840	B082	−7980	39800	80400	E016	−7933
30200	62280	C 76	−7745	40784	79260	E018	−7943
31730	63000	C 78	−7761	41780	76800	E020	−7966
33500	63720	C 80	−7793	42740	75120	E022	−7982
36680	65160	C 84	−7888	43730	74610	E024	−7984
38300	65790	C 86	−7927	44660	73200	E026	−7936
39830	66540	C 88	−7928	45680	71700	E028	−7897
41420	67200	C 90	−7923	46760	70200	E030	−7871
43010	67920	C 92	−7907	47600	68790	E032	−7835
44690	68580	C 94	−7879	48500	67380	E034	−7789
46250	69300	C 96	−7864	49520	65940	E036	−7738
48020	70020	C 98	−7847	42320	89100	F192	−8165
49400	70680	C100	−7836	43700	88080	F194	−8100
30200	76500	D546	−7897	45080	87000	F196	−8069
31640	75420	D548	−7868	46520	86010	F198	−8085
33080	74280	D550	−7833	47900	85050	F200	−8107
34520	73260	D552	−7791	49400	84000	F202	−8145
35900	72120	D554	−7788				

areas indicate regions where the TSLT (1) was not deposited, as it laps onto the top of the upper Lamont at the line labeled BASELAP, and (2) where it is truncated by UNCF, marked by the line TRUNCATE. Figure 7.3 shows a structure map of the top of the upper Lamont (TULM) that is truncated by the unconformity.

Figure B.4A shows an isochore map of the channel; the well-developed zero-line results from the use of additional positive and negative control points. Figure B.4B shows a structure map of the base of the channel.

TABLE B.4
Merged Well and Seismic Data

X	Y	ID	UNCF	TACK	TSLT	TULM	TLLM	CHNL	BLLM
32000	87015	W 1	−7888	−8138	−8139	−8153	−8158		−8210
38600	87030	W 2	−8020	−8058	−8073	−8078	−8086		−8137
44000	86400	W 3	−7943		−7957	−7958	−7970		−8025
48200	87060	W 4	−8003						
30650	82200	W 5	−7918	−8003	−8013	−8038	−8048	−8103	
35030	80901	W 6	−7905		−7943	−7964	−7976		−8027
39476	80970	W 7	−7900		−7928	−7935	−7949		−8004
44198	80910	W 8	−7998						
33530	76230	W 9	−7879		−7901	−7929	−7944	−8004	
38024	76170	W 10	−7817	−7878	−7933	−7959	−7976		−8032
42440	76650	W 11	−8018						
46955	76620	W 12	−7890						
30350	71700	W 13	−7903	−7907	−7961	−7963	−7980		−8033
35195	71745	W 14	−7815	−7946	−7998	−8025	−8045	−8120	
39650	71760	W 15	−7966	−7996	−8043	−8070	−8092		−8153
44180	71766	W 16	−7913		−7925	−7928	−7951		−8028
48800	70215	W 17	−7818	−7920	−7972	−7973	−7999		−8050
29240	81960	A610	−7920						
30800	82800	A612							
32270	83760	A614	−7923						
33770	84720	A616	−7918						
35240	85590	A618	−7927						
36740	86520	A620	−7969						
38210	87480	A622							
39680	88350	A624	−8043						
41240	89250	A626	−8059						
29840	67560	B068	−7932						
31280	66600	B070	−7883						
32720	65610	B072	−7842						
34190	64650	B074	−7842						
35720	63720	B076	−7874						
37100	62700	B078	−7871						
38600	61800	B080	−7861						
40100	60840	B082	−7875						
30200	62280	C 76	−7678						
31730	63000	C 78	−7740						
33500	63720	C 80	−7799						
36680	65160	C 84	−7902						
38300	65790	C 86	−7943						
39830	66540	C 88	−7962						
41420	67200	C 90	−7967						
43010	67920	C 92	−7931						
44690	68580	C 94	−7876						
46250	69300	C 96	−7854						
48020	70020	C 98							

(continued)

TABLE B.4
(continued)

X	Y	ID	UNCF	TACK	TSLT	TULM	TLLM	CHNL	BLLM
49400	70680	C100							
30200	76500	D546	−7849						
31640	75420	D548	−7880						
33080	74280	D550	−7869						
34520	73260	D552	−7820						
35900	72120	D554							
37280	71070	D556	−7892						
38660	70020	D558	−7965						
40040	69000	D560	−8008						
41450	67890	D562	−7986						
42800	66900	D564	−7922						
44240	65730	D566	−7852						
45500	64710	D568	−7773						
46940	63600	D570	−7693						
48200	62520	D572	−7632						
49610	61410	D574	−7567						
33800	89220	E003	−7896						
34880	87720	E005	−7915						
35840	86280	E007	−7943						
37040	84720	E010	−7948						
37940	83160	E012	−7928						
38840	81900	E014							
39800	80400	E016							
40784	79260	E018	−7923						
41780	76800	E020							
42740	75120	E022	−8004						
43730	74610	E024	−8001						
44660	73200	E026	−7927						
45680	71700	E028	−7877						
46760	70200	E030	−7857						
47600	68790	E032	−7824						
48500	67380	E034	−7768						
49520	65940	E036	−7696						
42320	89100	F192	−8032						
43700	88080	F194	−7971						
45080	87000	F196	−7946						
46520	86010	F198	−7973						
47900	85050	F200	−8003						
49400	84000	F202	−8032						

TABLE B.5
Points Added to Control Channel

X	Y	ID	THK
29000	60000	C 1	−80
29000	70000	C 2	−40
29000	80000	C 3	0
29000	90000	C 4	−50
40000	60000	C 5	−10
40000	70000	C 6	−60
40000	80000	C 7	−100
40000	90000	C 8	−180
36500	60000	C 9	55

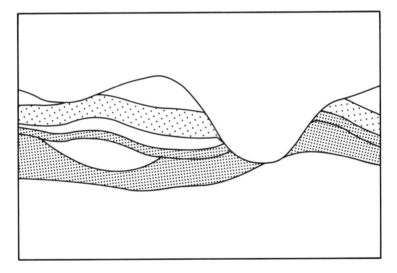

Figure B.2 Cross section through the generated stratigraphic framework, extending from (X = 29000, Y = 75000) to (X = 50000, Y = 75000).

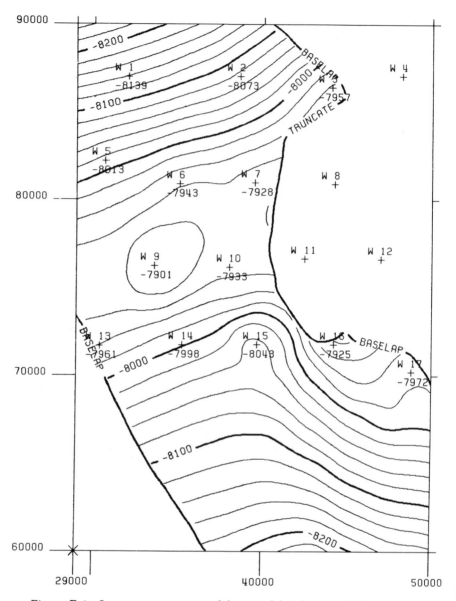

Figure B.3 Structure contour map of the top of the siltstone (TSLT), showing subcrops due to erosion (TRUNCATE) and nondeposition (BASELAP).

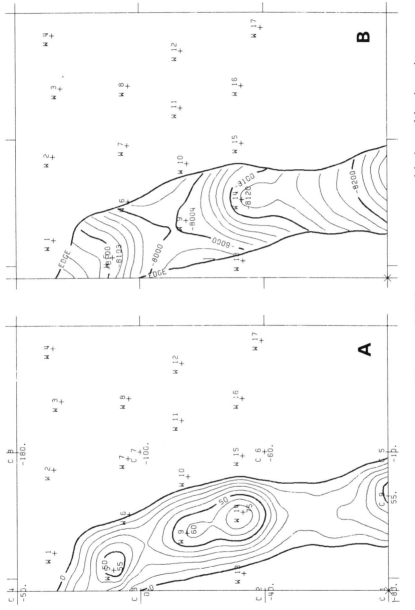

Figure B.4 (A) Isochore map of the channel. (B) Structure contour map of the base of the channel.

297

References

Agterberg, F. P., 1974, *Geomathematics: Mathematical Background and Geo-science Applications,* Elsevier, New York, 596p.

Ahlberg, J. H., E. W. Nilson, and J. L. Walsh, 1967, *The Theory of Splines and Its Applications,* Academic Press, New York, 280p.

Allen, D. R., 1964, Methods used in computing secondary equities, Wilmington field, Los Angeles County, California, Am. Assoc. Petroleum Geologists Bull. **48:**1150-1163.

Appleby, T. J., S. K. Jones, and T. M. Murin, 1984, Restored structure aids reservoir sandstone projection, *Oil & Gas Jour.* **82:**154-157.

Attoh, K., and E. H. T. Whitten, 1979, Computer programs for regression models for discontinuous structural surfaces, *Computers & Geosciences* **5:**47-71.

Badgley, P. C., 1959, *Structural Methods for the Exploration Geologist,* Harper & Row, New York, 280p.

Bates, R. L., and J. A. Jackson, 1980, *Glossary of Geology,* 2nd ed., American Geological Institute, Falls Church, Va., 758p.

Berg, R. R., 1975, Capillary pressure in stratigraphic traps, Am. Assoc. Petroleum Geologists Bull. **59:**939-956.

Bhattacharyya, B. K., 1965, Two-dimensional harmonic analysis as a tool for magnetic interpretation, *Geophysics* **30:**829-857.

Bhattacharyya, B. K., 1969, Bicubic spline interpolation as a method for treatment of potential field data, *Geophysics* **34:**402-423.

Billings, M. P., 1972, *Structural Geology,* 3rd ed., Prentice-Hall, Englewood Cliffs, N.J., 606p.

Bishop, M. S., 1960, *Subsurface Mapping,* Wiley, New York, 198p.

Bishop, R. S., 1979, Calculated compaction states of thick abnormally pressured shales, Am. Assoc. Petroleum Geologists Bull. **63:**918-933.

Bliss, J. D., and A. Rapport, 1983, GEOTHERM: The U. S. Geological Survey geothermal information system, *Computers & Geosciences* **9:**35-39.

Bolondi, G., F. Rocca, and S. Zanoletti, 1976, Automatic contouring of faulted surfaces, *Geophysics* **41:**1377-1393.

299

Bornemann, E., and J. H. Doveton, 1983, Lithofacies mapping of Viola limestone in South-Central Kansas, based on wireline logs, *Am. Assoc. Petroleum Geologists Bull.* **67**:609-623.

Bornhauser, M., 1958, Gulf Coast tectonics, *Am. Assoc. Petroleum Geologists Bull.* **42**:339-370.

Bowyer, A., 1981, Computing Dirichlet tessellations, *Computer Jour.* **24**:162-166.

Briggs, I. C., 1974, Machine contouring using minimum curvature, *Geophysics* **39**:39-48.

Brunet, M. F., 1984, Subsidence history of the Aquitaine Basin determined from subsidence curves, *Geol. Mag.* **121**:421-428.

Bugry, R., 1981, Computer contouring packages: An historical view, *Canadian Petrol. Geology Bull.* **29**:209-214.

Burkard, R. K., 1962, *Geodesy for the Layman*, U. S. Aeronautical Chart and Information Center, St. Louis, Mo., 64p.

Campbell, R. S., 1974, A generalized stepwise trend surface program—GSTR-1, *Computer Applications* **1**:161-170.

Cercone, K. R., 1983, Thermal history of Michigan Basin, *Am. Assoc. Petroleum Geologists Bull.* **67**:130-136.

Chapman, R. P., 1975, Data processing requirements and visual representation for stream sediment exploration geochemical surveys, *Jour. Geochem. Explor.* **4**:409-423.

Clark, D. A., 1981, A system for regional lithofacies mapping, *Canadian Petrol. Geology Bull.* **29**:197-208.

Clark, I., 1979, *Practical Geostatistics*, Applied Science Publishers, London, 129p.

Clark, N. J., 1960, *Elements of Petroleum Reservoirs*, Society of Petroleum Engineers, Dallas, Tex., 250p.

Clark, N. J., and W. P. Schultz, 1956, The analysis of problem wells, *Petroleum Engineer* **28**:B30-B38.

Clement, W. G., 1973, Basic principles of two-dimensional digital filtering, *Geophysical Prospecting* **21**:125-145.

Cohen, C. R., 1981, Time and temperature in petroleum formation: Application of Lopatin's method to petroleum exploration: Discussion, *Am. Assoc. Petroleum Geologists Bull.* **65**:1647-1648.

Cole, A. J., 1969, *An Iterative Approach to the Fitting of Trend Surfaces*, Computer Contribution No. 37, Kansas Geological Survey, Lawrence, Kansas, 27p.

Coppock, J. T., 1975, Maps by line printer, in J. C. Davis and M. K. McCullagh, eds., *Display and Analysis of Spatial Data*, Wiley, New York, pp. 137-154.

Costantino, M., 1983, A computerized database for the mechanical properties of coal, *Computers & Geosciences* **9**:53-58.

Craft, B. C., and M. F. Hawkins, 1959, *Applied Petroleum Engineering*, Prentice-Hall, Englewood Cliffs, N.J., 434p.

Crain, I. K., 1970, Computer interpolation and contouring of two-dimensional data: A review, *Geoexploration* **8**:71-86.

Crosby, W. O., 1912, Dynamic relations and terminology of stratigraphic conformity and unconformity, *Jour. Geology* **20**:289-299.

Crowell, J. C., 1959, Problems of fault nomenclature, *Am. Assoc. Petroleum Geologists Bull.* **43**:2653-2674.

Currie, J. B., 1956, Role of concurrent deposition and deformation of sediments in development of salt-dome graben structures, *Am. Assoc. Petroleum Geologists Bull.* **40**:3-16.

Dahlberg, E. C., 1972, Aspects of unbiased and biased contouring of geologic data by human and machine operators (abs.), *Geol. Soc. America, 1972 Meeting Abstracts with Program* **4**:482-483.

Dahlberg, E. C., 1975, Relative effectiveness of geologists and computers in mapping potential hydrocarbon exploration targets, *Math. Geol.* **7**:373-394.

Dahlberg, E. C., 1982, *Applied Hydrodynamics in Petroleum Exploration*, Springer-Verlag, New York, 161p.

David, M., 1977, *Geostatistical Ore Reserve Estimation*, Elsevier, New York, 364p.

Davis, J. C., 1973, *Statistics and Data Analysis in Geology*, Wiley, New York, 550p.

Davis, J. C., 1976, Contouring algorithms, in R. T. AAngeenbrug, ed., *Proceedings of the Second International Symposium on Computer-Assisted Cartography*, American Congress on Surveying and Mapping, New York, pp. 352-359.

Davis, J. C., 1986, *Statistics and Data Analysis in Geology*, 2nd ed., Wiley, New York, 646p.

Davis, M. W., and P. G. Culhane, 1984, Contouring very large datasets using kriging, in G. Verly, M. David, A. G. Journel, and A. Marechal, eds., *Geostatistics for Natural Resources Characterization*, D. Reidel, Dordrecht, Holland, pp. 599-619.

Davis, M. W., and M. David, 1980, Generating bicubic spline coefficients on a large regular grid, *Computers & Geosciences* **6:**1-6.

Davis, M. W., and C. Grivet, 1984, Kriging in a global neighborhood, *Math. Geol.* **16:**249-265.

Davis, S. M., and R. J. M. DeWiest, 1966, *Hydrogeology*, Wiley, New York, 463p.

De Boor, C., 1962, Bicubic spline interpolation, *Jour. Math. Phys.* **41:**212-218.

Dennison, J. M., 1968, *Analysis of Geologic Structures*, W. W. Norton, New York, 209p.

Dickinson, G., 1954, Subsurface interpretation of intersecting faults and their effects upon stratigraphic horizons, *Am. Assoc. Petroleum Geologists Bull.* **38:**854-877.

Dobrin, M. B., 1960, *Introduction to Geophysical Interpretation*, 2nd ed., McGraw-Hill, New York, 446p.

Dole, W. E., and N. F. Jordan, 1978, Slope mapping, *Am. Assoc. Petroleum Geologists Bull.* **62:**2427-2440.

Dorn, M., 1983, The use of automatic digitizers in geodata processing, *Computers & Geosciences* **9:**345-350.

Doveton, J. H., and A. J. Parsley, 1970, Experimental evaluation of trend surface distortions induced by inadequate data-point distributions, *Inst. Mining and Metallurgy Trans., Sec. B*, **79:**B197-B208.

Draper, N. R., and H. Smith, 1966, *Applied Regression Analysis*, Wiley, New York, 407p.

Dubrule, O., 1983, Two methods with different objectives: Splines and kriging, *Math. Geol.* **15:**245-257.

Dubrule, O., 1984, Comparing splines and kriging, *Computers & Geosciences* **10:**327-338.

Dubrule, O., and C. Kostov, 1986, An interpolation method taking into account inequality constraints: I. Methodology, *Math. Geol.* **18:**33-51.

Esler, J. E., P. F. Smith, and J. C. Davis, 1968, *KWIKR8, a FORTRAN IV Program for Multiple Regression and Geologic Trend Analysis*, Computer Contribution No. 28, Kansas Geological Survey, Lawrence, Kansas, 31p.

Fontaine, D. A., 1985, Mapping techniques that pay, *Oil & Gas Jour.* **83:**146-147.

Forgotson, Jr., J. M., 1977, Computer in petroleum exploration, in L. W. LeRoy, D. O. LeRoy, and J. W. Raese, eds., *Subsurface Geology: Petroleum, Mining, Construction*, 4th ed., Colorado School of Mines Press, Golden, Colo., pp. 466-482.

Foster, M. R., W. R. Jines, and K. van der Weg, 1970, Statistical estimation of systematic errors at intersections of lines of aeromagnetic survey data, *Jour. Geophys. Res.* **75:**1507-1511.

Fuller, B. D., 1967, Two-dimensional frequency analysis and design of grid operators, *Mining Geophysics* **2:**619-625.

Furlong, K. P., and J. D. Edman, 1984, Graphic approach to determination of hydrocarbon maturation in overthrust terrains, *Am. Assoc. Petroleum Geologists Bull.* **68:**1818-1824.

Garrett, R. G., 1974, Field data acquisition methods for applied geochemical surveys at the Geological Survey of Canada, *Geological Survey of Canada Paper 74-52*, 36p.

Gonzalez-Casanova, P., and A. Alvarez, 1985, Splines in geophysics, *Geophysics* **50:**2831-2848.

Grant, F., 1957, A problem in the analysis of geophysical data, *Geophysics* **22**:309-344.

Grant, F. S., and G. F. West, 1965, *Interpretation Theory in Applied Geophysics*, McGraw-Hill, New York, 584p.

Graybill, F. A., 1961, *An Introduction to Linear Statistical Models*, McGraw-Hill, New York, 463p.

Guidish, T. M., C. G. St. C. Kendall, I. Lerche, D. J. Toth, and R. F. Yarzab, 1985, Basin evaluation using burial history calculations: An overview, *Am. Assoc. Petroleum Geologists Bull.* **69**:92-105.

Guptill, S. C., 1983, The role of digital cartographic data in the geosciences, *Computers & Geosciences* **9**:23-26.

Hage, G. L., 1983, KRS: A fast special-purpose database system, *Computers & Geosciences* **9**:41-52.

Handley, E. J., 1954, Contouring is important, *World Oil* **138**:106-107.

Harbaugh, J. W., J. H. Doveton, and J. C. Davis, 1977, *Probability Methods in Oil Exploration*, Wiley, New York, 269p.

Harbaugh, J. W., and F. W. Preston, 1965, Fourier series analysis in geology, *Symposium on Computers and Computer Applications in Mining and Exploration*, vol. 1, School of Mines, University of Arizona, Tucson, pp. R1-R46.

Hedberg, H. D., 1976, *International Stratigraphic Guide*, Wiley, New York, 200p.

Hessing, R. C., H. K. Lee, A. Pierce, and E. N. Powers, 1972, Automatic contouring using bicubic functions, *Geophysics* **37**:669-674.

Hintze, W. H., 1971, Depiction of faults on stratigraphic isopach maps, *Am. Assoc. Petroleum Geologists Bull.* **55**:871-879.

Hitchon, B., 1984, Geothermal gradients, hydrodynamics, and hydrocarbon occurrences, Alberta, Canada, *Am. Assoc. Petroleum Geologists Bull.* **68**:713-743.

Hittelman, A. M., and D. R. Metzger, 1983, Marine geophysics: Database management and supportive graphics, *Computers & Geosciences* **9**:27-33.

Hobbs, B. E., W. D. Means, and P. F. Williams, 1976, *Structural Geology*, Wiley, New York, 571p.

Hollister, J. C., and T. L. Davis, 1977, Seismic prospecting for petroleum, in L. W. LeRoy, D. O. LeRoy, and J. W. Raese, eds., *Subsurface Geology: Petroleum, Mining, Construction*, 4th ed., Colorado School of Mines Press, Golden, Colo., pp. 425-437.

Horowitz, D. H., 1976, Mathematical modeling of sediment accumulation in prograding deltaic systems, in D. F. Merriam, ed., *Quantitative Techniques for Analysis of Sediments*, Pergamon Press, New York, pp. 105-119.

Houseknecht, D. W., 1984, Influence of grain size and temperature on intergranular pressure solution, quartz cementation, and porosity in a quartzose sandstone, *Jour. Sed. Petrol.* **54**:348-361.

Hubbert, M. K., 1953, Entrapment of petroleum under hydrodynamic conditions, *Am. Assoc. Petroleum Geologists Bull.* **37**:1954-2026.

Hubbert, M. K., 1967, Application of hydrodynamics to oil exploration, in *7th World Petroleum Congress Mexico City, Proceedings*, vol. 1B, pp. 59-75.

Hubral, P., and T. Krey, 1980, *Interval Velocities from Seismic Reflection Time Measurements*, Soc. Exploration Geophysicists, Tulsa, Okla., 203p.

Hunt, J. M., 1979, *Petroleum Geochemistry and Geology*, W. H. Freeman, San Francisco, 617p.

Iglehart, C. F., 1970, Descriptive classification of subsurface correlative tops, *Am. Assoc. Petroleum Geologists Bull.* **54**:1697-1705.

Iglehart, C. F., 1981, Computer applications committee cites data sources, *Am. Assoc. Petroleum Geologists Explorer*, December 1981, pp. 10-11.

Iglehart, C. F., 1982, Computer applications committee cites "tops" sources, *Am. Assoc. Petroleum Geologists Explorer*, January 1982, pp. 16-18.

James, W. R., 1966, FORTRAN IV Program Using Double Fourier Series for Surface Fitting of Irregularly Spaced Data, Computer Contribution No. 5, Kansas Geological Survey, Lawrence, Kansas, 19p.

James, W. R., 1967, Nonlinear models for trend analysis in geology, in D. F. Merriam and N. C. Cocke, eds., Computer Applications in the Earth Sciences: A Colloquium on Trend Analysis, Computer Contribution No. 12, Kansas Geological Survey, Lawrence, Kansas, pp. 26-30.

James, W. R., 1968, Development and Application of Nonlinear Regression Models in Geology, unpub. Ph.D. dissertation, Northwestern University.

James, W. R., 1970, Regression models for faulted structural surfaces, Am. Assoc. Petroleum Geologists Bull. **54:**638-646.

Johnson, C. R., 1977, Alternate geologic interpretation—Potential use of computer mapping in field development (abs.), Am. Assoc. Petroleum Geologists Bull. **61:**800.

Jones, T. A., 1984, Problems in using geostatistics for petroleum applications, in G. Verly, M. David, A. G. Journel, and A. Marechal, eds., Geostatistics for Natural Resources Characterization, D. Reidel, Dordrecht, Holland, pp. 651-667.

Jones, T. A., R. A. Baker, and W. H. Dumay, 1976, Executive system concept for processing geological data, Computers & Geosciences **2:**351-355.

Jones, T. A., and C. R. Johnson, 1983, Stratigraphic relationships and geologic history depicted by computer mapping, Am. Assoc. Petroleum Geologists Bull. **67:**1415-1421.

Jones, T. A., and N. F. Jordan, 1975, Structural mapping with data containing points that do not intersect the horizon of interest: An automated procedure (abs.), Am. Assoc. Petroleum Geologists & Soc. Econ. Paleontologists and Mineralogists Annual Meeting Abstracts **2**, pp. 40-41.

Journel, A. G., and Ch. J. Huijbregts, 1978, Mining Geostatistics, Academic Press, New York, 600p.

Junkins, J. L., G. W. Miller, and J. R. Jancaitis, 1975, A weighting function approach to modeling of irregular surfaces, Jour. Geophys. Res. **78:**1794-1803.

Kane, V. E., C. L. Begovich, T. R. Butz, and D. E. Myers, 1982, Interpretation of regional geochemistry using optimal interpolation parameters, Computers & Geosciences **8:**117-135.

Kostov, C., and O. Dubrule, 1986, An interpolation method taking into account inequality constraints: II. Practical approach, Math. Geol. **18:**53-73.

Krumbein, W. C., 1959, Trend-surface analysis of contour-type maps with irregular control-point spacing, Jour. Geophys. Res. **64:**823-834.

Krumbein, W. C., and F. A. Graybill, 1965, Statistical Models in Geology, McGraw-Hill, New York, 475p.

Krumbein, W. C., and L. L. Sloss, 1963, Stratigraphy and Sedimentation, 2nd ed., W. H. Freeman, San Francisco, 660p.

Langstaff, C. F., and D. Morrill, 1981, Geologic Cross Sections, International Human Resources Development Corporation, Boston, 108p.

Lee, W., 1955, Thickness maps can reveal mid-continent structure, World Oil **141:**77-82.

Lee, Y. W., 1960, Statistical Theory of Communication, Wiley, New York, 251p.

Leet, L. D., S. Judson, and M. E. Kauffman, 1982, Physical Geology, 6th ed., Prentice-Hall, Englewood Cliffs, N.J., 487p.

Levorsen, A. I., 1967, Geology of Petroleum, 2nd ed., W. H. Freeman, San Francisco, 724p.

Low, J. W., 1977, Subsurface maps and illustrations, in L. W. LeRoy, D. O. LeRoy, and J. W. Raese, eds., Subsurface Geology: Petroleum, Mining, Construction, 4th ed., Colorado School of Mines Press, Golden, Colo., pp. 244-284.

Mandelbaum, H., 1963, Statistical and geological implications of trend mapping with non-orthogonal polynomials, Jour. Geophys. Res. **68:**505-519.

Marechal, A., 1984, Kriging seismic data in presence of faults, in G. Verly, M. David, A. G. Journel, and A. Marechal, eds., Geostatics for Natural Resources Characterization, D. Reidel, Dordrecht, Holland, pp. 271-294.

Maslyn, R. M., and F. J. Phillips, 1984, Computer modeling of Minnelusa (Pennsylvanian-Permian) paleotopography in eastern Powder River Basin, Wyoming, with a case history (abs.), *Am. Assoc. Petroleum Geologists Bull.* **68**:503-504.

Mather, P. M., 1976, *Computational Methods of Multivariate Analysis in Physical Geography,* Wiley, New York, 532p.

McCammon, R. B., 1974, The statistical treatment of geochemical data, in A. A. Levinson, ed., *Introduction to Exploration Geochemistry,* Applied Publishing, Wilmette, Ill., pp. 469-508 (1980 Supplement, 2nd ed., pp. 835-843).

McLain, D. H., 1974, Drawing contours from arbitrary data points, *Computer Jour.* **17**:318-324.

McLain, D. H., 1976, Two-dimensional interpretation from random data, *Computer Jour.* **19**:178-181.

Merriam, D. F., and N. C. Cocke, eds., 1967, *Computer Applications in the Earth Sciences: A Colloquium on Trend Analysis,* Computer Contribution No. 12, Kansas Geological Survey, Lawrence, Kansas, 62p.

Mesko, A., 1984, *Digital Filtering: Applications in Geophysical Exploration for Oil,* Wiley, New York, 512p.

Middleton, M. F., and D. A. Falvey, 1983, Maturation modeling in Otway Basin, Australia, *Am. Assoc. Petroleum Geologists Bull.* **67**:271-279.

Miesch, A. T., and J. J. Connor, 1968, *Stepwise Regression and Nonpolynomial Models in Trend Analysis,* Computer Contribution No. 27, Kansas Geological Survey, Lawrence, Kansas, 40p.

Moore, C. A., 1963, *Handbook of Subsurface Geology,* Harper & Row, New York, 235p.

Mosteller, F., and J. W. Tukey, 1977, *Data Analysis and Regression: A Second Course in Statistics,* Addison-Wesley, Reading, Mass., 588p.

Nordstrom, J. E., 1985, An algorithm for contouring geologic spatial data with known discontinuities, *The Compass* **62**:74-89.

Ocamb, R. D., 1961, Growth faults of south Louisiana, *Gulf Coast Assoc. Geol. Societies Trans.* **11**:139-176.

Patnode, H. W., and R. A. Hodgson, 1964, Three-dimensional geologic maps, *Amer. Jour. Sci.* **262**:274-278.

Paynter, D. D., 1970, Main Pass Block 35 field, Louisiana: Paleostructural analysis, *Am. Assoc. Petroleum Geologists Bull.* **54**:783-788.

Peikert, E. W., 1970, Interactive computer graphics and the fault problem (abs.), *Am. Assoc. Petroleum Geologists Bull.* **54**:556.

Pennebaker, P. E., 1972, Vertical net sandstone determination for isopach mapping of hydrocarbon reservoirs, *Am. Assoc. Petroleum Geologists Bull.* **56**:1520-1529.

Perrier, R., and J. Quiblier, 1974, Thickness changes in sedimentary layers during compaction history: Methods for quantitative evaluation, *Am. Assoc. Petroleum Geologists Bull.* **58**:507-520.

Peucker, T. K., M. Tichenor, and W. D. Rose, 1975, The computer version of three relief representations, in J. C. Davis and M. K. McCullagh, eds., *Display and Analysis of Spatial Data,* Wiley, New York, pp. 187-197.

Pfaltz, J. L., 1975, Representation of geographic surfaces within a computer, in J. C. Davis and M. K. McCullagh, eds., *Display and Analysis of Spatial Data,* Wiley, New York, pp. 210-230.

Philip, G. M., and D. F. Watson, 1982, A precise method for determining contoured surfaces, *Australian Petrol. Explor. Assoc. Jour.* **22**:205-212.

Pouzet, J., 1980, Estimation of a surface with known discontinuities for automatic contouring purposes, *Math. Geol.* **12**:559-575.

Powers, E. N., 1984, Should a microcomputer be used to make a geologic contour map? (abs.), *Am. Assoc. Petroleum Geologists Bull.* **68**:519.

Press, F., and R. Siever, 1982, *Earth,* 3rd ed., W. H. Freeman, San Francisco, 613p.

Preston, F. W., and J. W. Harbaugh, 1965, *BALGOL Program and Geologic Application for Single and Double Fourier Series Using IBM 7090/94 Computers*, Special Distribution Publication No. 24, Kansas Geological Survey, Lawrence, Kansas, 72p.

Ragan, D. M., 1985, *Structural Geology*, 3rd ed., Wiley, New York, 393p.

Rasmussen, K. L., and P. V. Sharma, 1979, Bicubic spline interpolation: A quantitative test of accuracy and efficiency, *Geophys. Prospecting* **27**:394-408.

Reiter, W. A., 1947, Contouring fault planes, *World Oil* **126**:34-35.

Rhind, D. W., 1971, Automated contouring—An empirical evaluation of some differing techniques, *Cartographic Jour.* **8**:145-158.

Rich, J. L., 1935, Graphical method for eliminating regional dip, *Am. Assoc. Petroleum Geologists Bull.* **19**:1538-1543.

Richardus, P., and R. K. Adler, 1972, *Map Projections for Geodesists, Cartographers, and Geographers*, American Elsevier, New York, 174p.

Ripley, B. D., 1981, *Spatial Statistics*, Wiley, New York, 252p.

Rittenhouse, G., 1971a, Pore-space reduction by solution and cementation, *Am. Assoc. Petroleum Geologists Bull.* **55**:80-91.

Rittenhouse, G., 1971b, Compaction of sands containing different percentages of ductile grains: A theoretical approach, *Am. Assoc. Petroleum Geologists Bull.* **55**:92-96.

Roberts, J. L., 1982, *Introduction to Geological Maps and Structures*, Pergamon Press, New York, 332p.

Robinson, J. E., 1969, Spatial filters for geological data, *Oil & Gas Jour.* **67**:132-140.

Robinson, J. E., 1982, *Computer Applications in Petroleum Geology*, Hutchinson Ross, Stroudsburg, Pa., 164p.

Robinson, J. E., H. A. K. Charlesworth, and M. J. Ellis, 1969, Structural analysis using spatial filtering in interior plains of south-central Alberta, *Am. Assoc. Petroleum Geologists Bull.* **53**:2341-2367.

Robinson, J. E., and D. F. Merriam, 1971, Z-trend maps for quick geologic recognition of geologic patterns, *Math. Geol.* **3**:171-181.

Robinson, J. E., and D. F. Merriam, 1972, Enhancement of patterns in geologic data by spatial filtering, *Jour. Geol.* **80**:333-345.

Robinson, J. E., and D. F. Merriam, 1984, Computer evaluation of prospective petroleum areas, *Oil & Gas Jour.* **82**:135-138.

Rose, A. W., 1972, Statistical interpretation techniques in geochemical exploration, *Soc. Mining Engineers (AIME) Trans.* **252**:233-239.

Russell, W. L., 1955, *Structural Geology for Petroleum Geologists*, Maple Press, York, Pa., 427p.

Sampson, R. J., 1975, The SURFACE II graphics system, in J. C. Davis and M. K. McCullagh, eds., *Display and Analysis of Spatial Data*, Wiley, New York, pp. 244-266.

Savoy, B. V., and A. L. Valentine, 1961, How to convert bed-thickness data from directional wells, *Oil & Gas Jour.* **59**:68-72.

Schmidt, A. H., and W. A. Zafft, 1975, Programs of the Harvard University Laboratory for Computer Graphics and Spatial Analysis, in J. C. Davis and M. K. McCullagh, eds., *Display and Analysis of Spatial Data*, Wiley, New York, pp. 231-243.

Schmoker, J. W., 1984, Empirical relation between carbonate porosity and thermal maturity: An approach to regional porosity prediction, *Am. Assoc. Petroleum Geologists Bull.* **68**:1697-1703.

Sebring, Jr., L., 1958, Chief tool of the petroleum exploration geologist: The subsurface structure map, *Am. Assoc. Petroleum Geologists Bull.* **42**:561-587.

Shaw, B. R., 1977, Evaluation of distortion of residuals in trend surface analysis by clustered data, *Math. Geol.* **9**:507-517.

Sibson, R., and G. D. Thomson, 1981, A seamed quadratic element for contouring, *Computer Jour.* **24**:378-382.

Siever, R., 1983, Burial history and diagenetic reaction kinetics, *Am. Assoc. Petroleum Geologists Bull.* **67**:684-691.

Silver, B. A., 1983, *Subsurface Exploration Stratigraphy*, Institute of Energy Development, Oklahoma City, 342p.

Sloss, L. L., and W. Scherer, 1975, Geometry of sedimentary basins: Applications to Devonian of North America and Europe, *Geol. Soc. America Mem.* **142**:71-88.

Spector, A., and F. S. Grant, 1970, Statistical models for interpreting aeromagnetic data, *Geophysics* **35**:293-302.

Spencer, E. W., 1977, *Introduction to the Structure of the Earth*, 2nd ed., McGraw-Hill, New York, 640p.

Sprunt, B. F., 1975, Relief representation in automated cartography: an algorithmic approach, in J. C. Davis and M. K. McCullagh, eds., *Display and Analysis of Spatial Data*, Wiley, New York, pp. 173-186.

Stephenson, L. P., 1977, Porosity dependence on temperature: Limits on possible effect, *Am. Assoc. Petroleum Geologists Bull.* **61**:407-415.

Sutcliffe, D. C., 1976, A remark on a contouring algorithm, *Computer Jour.* **19**:333-335.

Swain, C. J., 1976, A FORTRAN IV program for interpolating irregularly spaced data using difference equations for minimum curvature, *Computers & Geosciences* **1**:231-240.

Switzer, P., 1965, Reconstructing patterns from sample data, *Annals Math. Statistics* **36**:138-154.

Szumilas, D., 1977, Using computer for time-to-depth conversion and structure mapping in complexly faulted areas (abs.), *Am. Assoc. Petroleum Geologists Bull.* **61**:834-835.

Tanaka, K., 1950, The relief contour method of representing topography on maps, *Geographical Rev.* **40**:444-456.

Tissot, B. P., 1984, Recent advances in petroleum geochemistry applied to hydrocarbon exploration, *Am. Assoc. Petroleum Geologists Bull.* **68**:545-563.

Trump, R. P., and H. W. Patnode, 1960, Graphic stereoscopic representation of irregular surfaces, *Am. Assoc. Petroleum Geologists Bull.* **44**:1570-1571.

Vail, P. R., R. M. Mitchum, R. G. Todd, J. M. Widmier, S. Thompson, J. B. Sangree, J. N. Bubb, and W. J. Hatelid, 1977, Seismic stratigraphy and global changes of sea level, in C. E. Payton, ed., *Seismic Stratigraphy – Applications to Hydrocarbon Exploration*, Memoir 26, American Association of Petroleum Geologists, Tulsa, Okla., pp. 49-212.

van Hinte, J. E., 1978, Geohistory analysis – applications of micropaleontology in exploration strategy, *Am. Assoc. Petroleum Geologists Bull.* **62**:201-222.

Walters, R. F., 1969, Contouring by machine: A user's guide, *Am. Assoc. Petroleum Geologists Bull.* **53**:2324-2340.

Waples, D. W., 1980, Time and temperature in petroleum formation: Application of Lopatin's method to petroleum exploration, *Am. Assoc. Petroleum Geologists Bull.* **64**:916-926.

Waters, K. H., 1981, *Reflection Seismology: A Tool for Energy Resources Exploration*, Wiley, New York, 453p.

Waters, N. M., 1981, Computer mapping: A review of what is available and what is useful for exploration purposes, *Canadian Petroleum Geol. Bull.* **29**:182-196.

Watson, D. F., 1981, Computing the n-dimensional Delauney tesselation with application to Voronoi polytopes, *Computer Jour.* **24**:161-172.

Watson, D. F., 1982, ACORD: Automatic contouring of raw data, *Computers & Geosciences* **8**:97-101.

Watson, D. F., and G. M. Philip, 1984, Triangle-based interpolation, *Math. Geol.* **16**:779-795.

Watson, G. S., 1971, Trend-surface analysis, *Math. Geol.* **3**:215-226.

Watson, G. S., 1984, Smoothing and interpolation by kriging and with splines, *Math. Geol.* **16**:601-615.

Wharton, Jr., J. B., 1948, Isopachous maps of sand reservoirs, *Am. Assoc. Petroleum Geologists Bull.* **32**:1331-1339.

Whitten, E. H. T., 1970, Orthogonal polynomial trend surfaces for irregularly-spaced data, *Math. Geol.* **2:**141-152 (erratum in *Math. Geol.* **3:**329-330, 1971).

Whitten, E. H. T., 1972, More on "Orthogonal polynomial trend surfaces for irregularly-spaced data," *Math. Geol.* **4:**83.

Whitten, E. H. T., 1974, Orthogonal-polynomial contoured trend-surface maps for irregularly-spaced data, *Computer Applications* **1:**171-183.

Whitten, E. H. T., and W. A. Beckman, Jr., 1969, Fold geometry within part of Michigan Basin, Michigan, *Am. Assoc. Petroleum Geologists Bull.* **53:**1043-1057.

Williams, J. D., 1984, Interactive flattening as an exploration tool (abs.), *Am. Assoc. Petroleum Geologists Bull.* **68:**540.

Wilson, M. D., 1975, Comparison of fan-pass spatial filtering and polynomial surface-fitting models for numerical map analysis, *Geol. Soc. America Mem.* **142:**351-366.

Wilson, Jr., C. W., and R. G. Stearns, 1963, Quantitative analysis of Ordovician and younger structural development of Nashville Dome, Tennessee, *Am. Assoc. Petroleum Geologists Bull.* **47:**823-832.

Wren, A. E., 1975, Contouring and the contour map: A new perspective, *Geophys. Prospecting* **23:**1-17.

Yoeli, P., 1967, The mechanization of analytical hill shading, *Cartographic Jour.* **4:**82-88.

Yoeli, P., 1975, Compilation of data for computer-assisted relief cartography, in J. C. Davis and M. K. McCullagh, eds., *Display and Analysis of Spatial Data*, Wiley, New York, pp. 352-367.

Zurfleuh, E. G., 1967, Applications of two-dimensional linear wavelength filtering, *Geophysics* **32:**1015-1035.

Index